WEATHER

A Collection of Rhymes, Poems, Folklore and Observations on Weather Forecasting.

By

Dr. Gerald A. Walford

Rhymes and reasons

For the seasons

Copyright

This book, "WEATHER

A Collection of Rhymes, Poems, Folklore and Observations on Weather Forecasting."

is copyright © 2018.

Also by Dr. Gerald A. Walford

GOLF: Class A – Professional Golfers' Association of America.
World Golf Teachers' Federation
The Golf Superbook.
Slapshot Golf: Learning Golf Through the Hockey Slapshot and Baseball Swing.
Golf's Power Secrets: How the Great One's Did it and Taught it.
The Golf Whisperer.
Performance Golf.
Golf Instantly Simple – An Accurate Method for Long and Straight Golf Shots and putting Strokes.
The Life of a Mental Patient on the PGA Golf Tour – I may be crazy but I am not Stupid. Sequels 1, 2, and 3.

SPORT
The Death of Human Sport.
Sport, Religion and War – Through the Ages.
Croquet – a Game for All

MENTAL HEALTH
Psychopaths, Sociopaths and Antisocial Behavior – Are You One.

NOVEL
The Lady Spy and the Con Man. Based on facts.
One Flew Over the Cuckoo Golf Green and Dropped His Fertilizer.

HISTORY
Elite Women in Dangerous Jobs – Pilots, Spies, Snipers, and Battlefield Nurses in WWII with pre and post history. (Color picture version)

Brave Women Doing Dangerous Jobs In WWII. Pilots, Spies, Snipers, And Battlefield Nurses (Black and White picture version).

Sport, Religion and War

HOCKEY
Coaching Hockey.
Hockey Skills.
Youth Hockey.

Ice Hockey: An Illustrated Guide for Coaches.
Coaching Girls Ice Hockey: For Ages 11 and Up.

HUMOR
Humor and Personality.
Sport Humor.
Humorous Stories and Sayings – Some true - Some Almost True –Some Just Funny.

BASEBALL SOFTBALL
The Baseball/Softball Swing of the Future.

CONCENTRATION
Concentration and Other Mental Skills for Sports, Life and the Arts.

LEADERSHIP
Determining the Future Through Leadership Skills.

TEACHING
The Tao of Teaching: The Teaching Experience.

BIOGRAPHY
Controlling Adversity.

Email: geraldwalford2@hotmail.com

TABLE OF CONTENTS

INTRODUCTION

Many of the rhymes and poems are in old English so sometimes there may be a little difficulty in understanding but the message will still come through.

In some cases, the weather predictions may prevail only to the local area where they originated. In some cases, the local area could not be determined.

The photos are in black and white because color photos would raise the retail price by three times.

WEATHER'S ENVIRONMENTAL CONDITION

http://www.42explore.com/weather.htm

Weather is the environmental condition of the air over the ground surface at a certain time and place.

The environmental conditions will range with temperature, moisture, winds and clouds.

One's location has a climate which is determined by the environmental factors of temperature, rainfall, wind and clouds over time, usually a long time.

Oceans, mountains, plains, surface terrain, etc. all will effect the weather.

Meteorology is a study of the weather. **Meteorologists** are the people who study the weather.

The forecast is a prediction of the weather into the short or long-term future.

Weather stations or equipment is placed around the world to obtain information as to the weather and its movement. Their instruments will measure everything about the weather like temperature, rainfall, wind speed, wind direction, air pressure, humidity, etc. This analysis is very scientific and actually quite accurate although sometimes the unexpected may happen.

Precipitation is water that falls from the sky as rain, sleet, hail, or snow. Winds occur when the air is heated by the sun and because the heating is uneven, winds develop. The earth's surface and terrain often determine the movement of the weather.

The **air** moves as currents over the earth surface. Air pressure is the weight, 14.7 pounds per square inch, of the air pressing down against the surface of the earth. This pressure changes from day to day by going up or down and is often an indication of the weather and future weather. A barometer measures the air pressure.

Relative humidity is the amount of water or moisture in the air and how much the air can hold without dropping to the earth as rain, snow, etc. When the air's relative humidity is 100%, the air releases the moisture and it drops. In arid areas, like deserts, the relative humidity may be a low as 5 or 10%.

Fronts are bodies of cold or warm air moving over the earth's surface. These fronts often interact against each other to produce favorable or unfavorable weather.

Although we have many tools to help us in weather forecasting, weather is still difficult to predict. Science has come a long way in weather forecasting. Over the years, rhymes, poems, and sayings have developed orally and past on as folklore. Quite often, they have had validation through science.

Thunderstorm

Weather determines life. Weather is essential to the farmer, to the traveler, to the type of attire or clothes one wears, sports and recreation is controlled by the weather, etc. It is almost fair to say whether determines everything.

Ever since man existed, he tried to predict the weather. It was not easy. Trends in the weather would bring about certain occurrences like sunshine, rain, winds, etc. These trends were maintained cognitively, by memory, and expressed, and past on by oral means because there was no writing back then. When writing was invented records were kept and their ability to predict became better simply by trial and error. This method lasted until the high tech age when science was able to predict the weather much more efficiently, maybe, but sometimes not as accurately as we may like.

It is interesting to note that some of the old folklore rhymes, poems, and word-of-mouth often proved fairly accurate as science soon proved. We must realize that the weather is simply not a chaotic pattern. Certain circumstances predict what is to follow. For instance, large Cumulus Nimbus clouds, especially with the thunderhead, usually mean rain. There is a pattern to the prediction that changes, sometimes very sudden alter the predictions.

Marestail shows moisture at high altitude, signalling the later arrival of wet weather.

https://en.wikipedia.org/wiki/Weather_lore

Cumulus humilis indicates a dry day ahead.

https://en.wikipedia.org/wiki/Weather_lore

The thermometer and the mercury barometer were early inventions into the science of weather prediction.

Native Americans used the *weather stick* for their predictions. The natives noticed the behavior of dry branches prior to the arrival of weather changes so they ingeniously developed the weather stick. This consisted of balsam or birch rod mounted outdoors that would twist upwards for low humidity and downwards for high humidity environments. This tool predated the mercury barometer.

Advancing monsoon clouds and showers in Aralvaimozhy, near Nagercoil, India.

The monsoons are an example of a consistent yearly weather pattern that comes in monsoon season.

CLOUDS

If clouds move against the wind, rain will follow.

This is a tricky and difficult one as it only occurs under special circumstances; otherwise, it is false.

Stand with one's back to the ground level wind and observe the clouds. In the northern hemisphere if the upper level clouds are moving from the right a low-pressure area has passed and the weather will be improving. If from the left a low-pressure area is arriving in the weather will deteriorate. In the southern hemisphere – reverse.

Source: List of cloud types - https://en.wikipedia.org

Source: File:Helkivad ööpilved Kuresoo kohal.jpg - https://en.wikipedia.org

"Noctilucent cloud over Estonia"

Source: File:Nacreous clouds Antarctica.jpg - https://en.wikipedia.org

Stratospheric nacreous clouds over Antarctica

Source: File:Cirrus clouds2.jpg - https://en.wikipedia.org

Cirrus uncinus clouds (V-2)

Source: File:CirrusField-color.jpg - https://en.wikipedia.org

Source: File:Cumulus clouds panorama.jpg - https://en.wikipedia.org

Cumuliform cloudscape over Swifts Creek, Australia.

FOLKLORE

Here are some sayings that seem to have been proven scientifically. To use color photos the retail price would triple so black and white photos will be used.

Red sky at night, with dust and clouds moving away to the west

https://en.wikipedia.org/wiki/Red_sky_at_morning

A red sunset probably means dry weather the next day.

https://en.wikipedia.org/wiki/Weather_lore

Red sky at morning, at sea

The common phrase **"red sky at morning"** is a line from an ancient rhyme often repeated by mariners:

Red sky at night, sailors' delight.
Red sky at morning, sailors take warning.

This same poem had its variations and one of the more common usage was shepherds instead of sailors.

Red sky at night, shepherds' delight;
Red sky in the morning, shepherds' warning.

Depending on the location and the people involved words other than sailors were more likely used, but the meanings are still the same. England has lots of sheep roaming the countryside so naturally the words shepherds would be common to the English. In the USA, very few sheep so sailors would seem more appropriate as sailors were popular on the East Coast.

This rhyme has persisted over two millenniums. In fact, in the Bible: Matthew 16. 2-b, Jesus says: *When it is evening, you say, "It will be fair weather; for the sky is red." And in the morning, "It will be stormy today, for the sky is red and threatening.*

"*Bible Gateway passage: Matthew 16:2-3 - New American Standard Bible*". Bible Gateway. Retrieved 2017-08-15.

When Clouds Look Like Black Smoke

When Clouds Look Like Black Smoke a Wise Man Will Put On His Cloak,

ATMOSPHERIC PRESSURE

Low pressure regions
https://en.wikipedia.org/wiki/Weather_lore

When the wind is blowing in the North
No fisherman should set forth,
When the wind is blowing in the East,
'Tis not fit for man nor beast,
When the wind is blowing in the South
It brings the food over the fish's mouth,
When the wind is blowing in the West,
That is when the fishing's best!

This poem was created mainly for the Western European seas when under a low-pressure. The northerly winds are cold, and blustery with heavy waves to create dangerous sailing. The Eastern winds are warm, dry and dusty during the summer and very cold during the winter. Eastern winds are very annoying and uncomfortable. The southern winds are usually warm and provide pleasant sailing and fishing weather although the fish may not necessarily be more abundant. The westerly wind brings fair and clear whether with a fairly consistent wind.

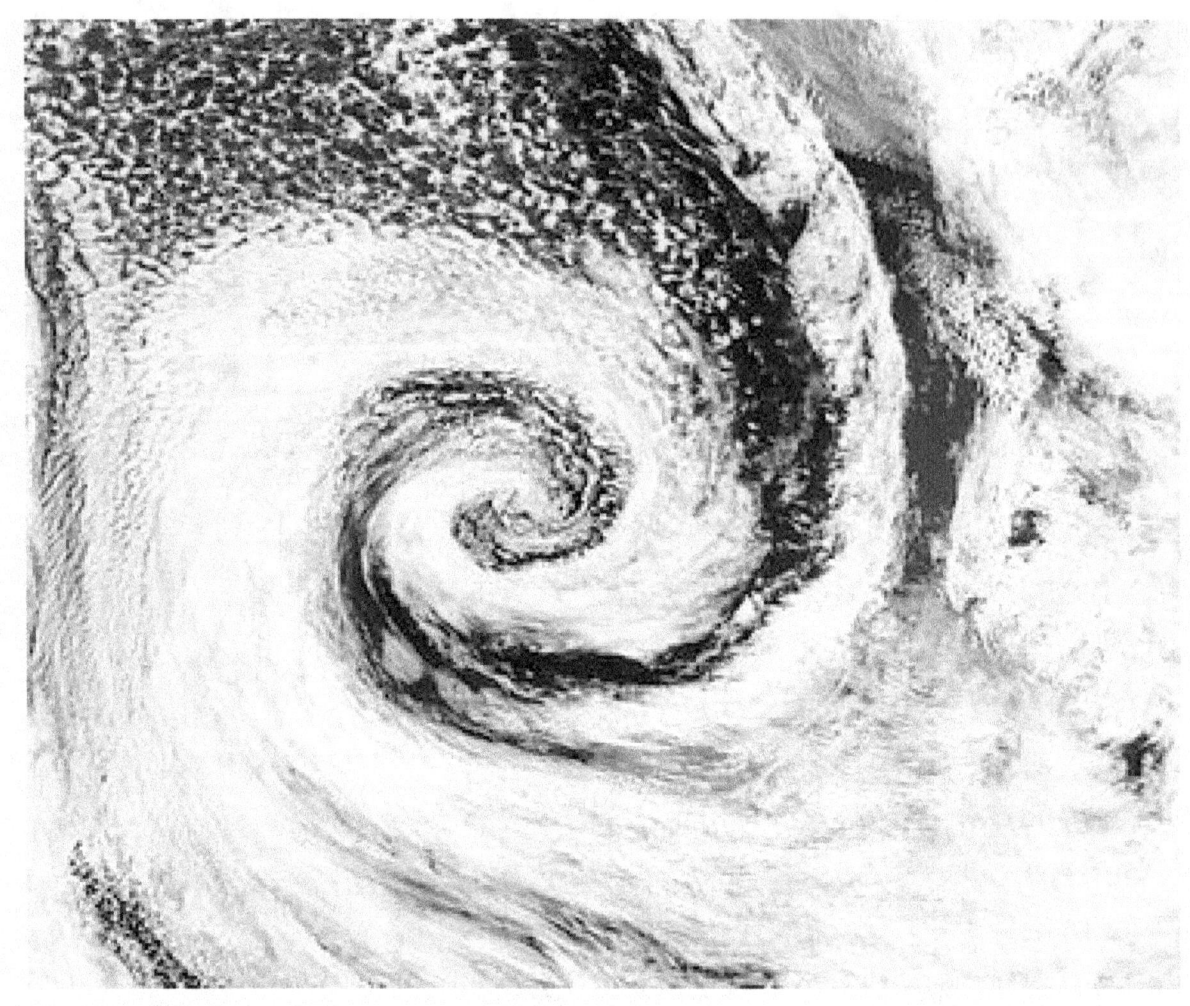

A clockwise spinning low pressure area or cyclone off southern Australia. The center of the spiral-shaped cloud system is also the center of a low, and usually is where the pressure is lowest.

This low-pressure system over **Iceland spins counter-clockwise due** to balance between the Coriolis force and the pressure gradient force and in the northern half of the world.

Calm conditions: high-pressure regions

High-pressure areas are usually indicative of calm conditions with clear skies. The air is descending so clouds, wind, precipitation are not evident.

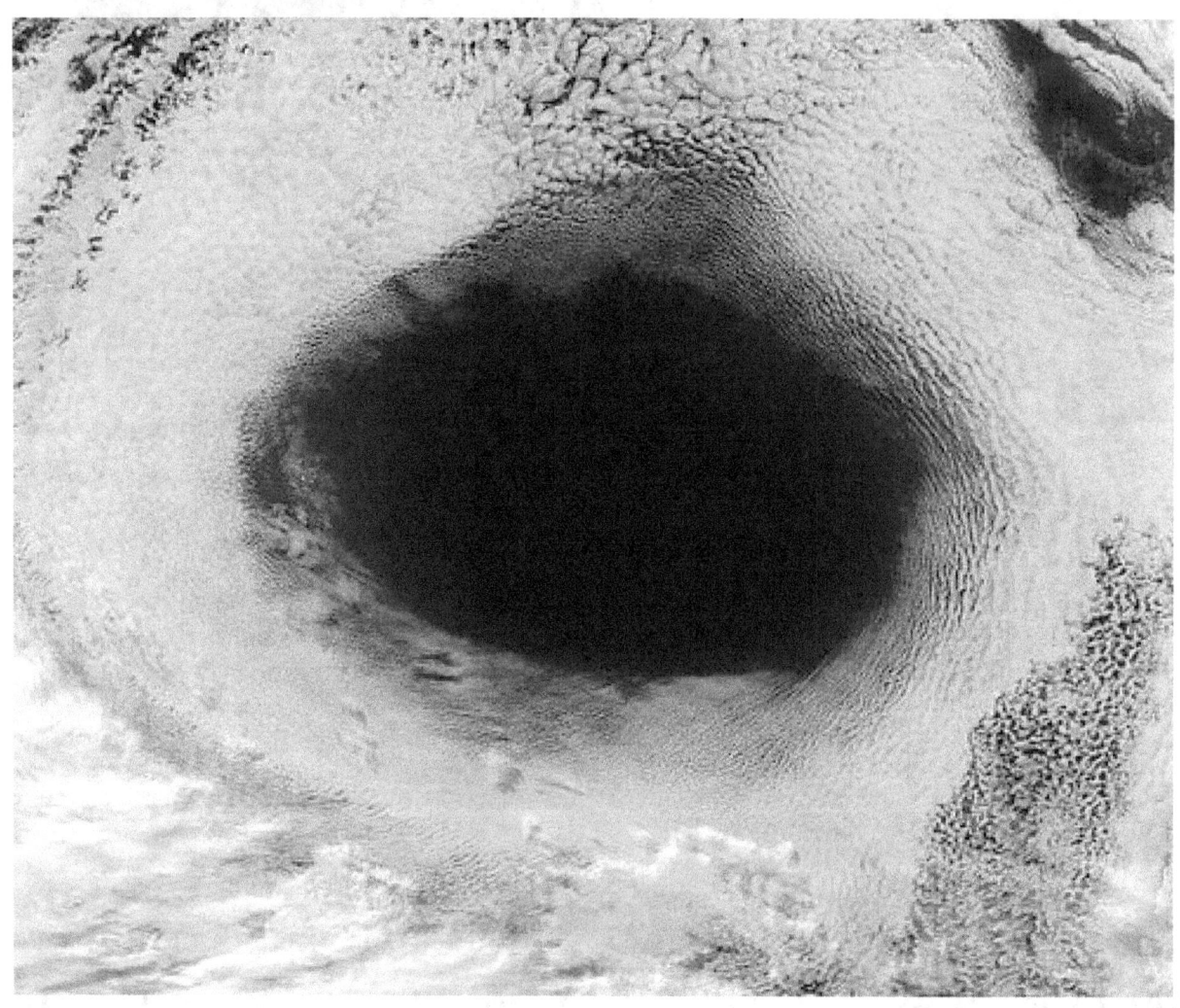

Satellite image of an unusual high-pressure area south of Australia, evidenced by the clearing in the clouds.

https://en.wikipedia.org/wiki/High-pressure_area

RING AROUND THE MOON

https://en.wikipedia.org/wiki/Weather_lore

When halo rings the moon or sun,
rain's approaching on the run.

Solar halo is precursor to rain.

Lunar corona

The halo around the sun or moon is caused but the refraction of the body's light on the ice crystals at high altitudes. This high-level moisture is a precursor or indication that moisture is moving into the area and active weather is on its way. Halos bring out a "milk sky" when the sky appears clear that the typical blue seems to have a washed out or barely noticeable blue color. This is an indicator of an approaching low. On the coldest days of winter is brings clear air and the possibility of "sun dogs" and a change of the weather.

Sun dog or **mock sun**, formally called a **parhelion** in meteorology, is an atmospheric optical phenomenon that consists of a bright spot to the left or right of the Sun. Two sun dogs often flank the Sun within a 22° halo. Very bright sun dogs in Fargo, North Dakota in the above photo.

HUMIDITY

When windows won't open, and the salt clogs the shaker,
The weather will favor the umbrella maker!

The high level of moisture in the air has a tendency to swell wood making doors and windows a little sticky and sometimes difficult to slide or move. Salt is a moisture absorber. Morton's salt has added magnesium carbonate or calcium silicate to iodized salt to prevent the salt from clomping together. This is why their logo has a girl with an umbrella and the slogan, "When it rains, it pours".

FOG

A summer fog for fair,
A winter fog for rain.
A fact most everywhere,
In valley or on plain.

Fog can be considered as simply a cloud of water droplets or ice crystals suspended in the air near the Earth's surface.

Fog is created when the air cools enough that the vapor pressure encourages condensation instead of evaporation.

Cool summer nights occur when the sky is clear and the heat radiates off the earth's surface into space. Cloudy skies act like a blanket in keeping the heat close to the Earth's surface so if it is cool enough and clear enough for fog to form there's a good chance the next day will be clear.

The air is typically more humid above the ocean, or a large lake, than over land. When this humid air moves over the cold land it will form fog and even precipitation.

To the east of North America's Great Lakes, this common phenomenon is known as "the lake effect".

"Ice fog" occurs in urban or dense populated areas when the air is so cold vapor pressure results in condensation (fog) and the additional vapor from automobiles, household furnaces, and industrial plants to create a heavy fog and perhaps sometimes referred to as smog because of man-made vapor.

SOUNDS

When sounds travel far and wide,
A stormy day will betide.

This poem is true during the summer only. Moist air conducts sound better than dry air and so sound carries further in the moist air.

In winter, temperature is a factor. If the air is warm and moist the sound travels farther. If the air is very cold, the air is now dense and also conducts sound very good. When sounds carry over a long distance the cold clear weather is likely to linger.

ACHES AND PAINS

A coming storm your shooting corns presage,
And aches will throb, your hollow tooth will rage.

Older people are familiar with this one and medical studies seem to support it. It seems that falling atmospheric pressure causes the blood vessels to slightly dilate and irritate the nerves near corns, cavities, and arthritic joints.

MONTHLY PREDICTIONS

February

GROUNDHOG DAY

Empirical studies have not found any hard-core evidence that if the groundhog sees its shadow on February 2, six weeks of winter remain.

February 2 seems to carry much folk lore as this day is also known as Candlemas, Brigid's Day, or St. Blaise's Day (St. Blaze's Day).

In Germany, *If the badger is in the sun at Candlemas, the badger returns to his lair for six weeks.*

In France, *If it is fair weather on Candlemas, the bear returns to its cave for six weeks.*

Or
if it is sunny on Candlemas the wolf returns to its cave for six weeks, and if not, for forty days.

In French Canada, it may be a marmot or groundhog (*siffleux*), bear, skunk, otter etc. which if it sees its shadow on Candlemas, causes winter to prolong for 40 days.

In England, the traditional weather lore is:

If Candlemas Day be fair and bright.

Winter will have another fight.

If Candlemas Day brings cloud and rain.

Winter won't come again.

MARCH

English proverb from the 19th century:

March comes in like a lion and goes out like a lamb:

Still common today although one wonders if anyone ever remembers to the end of the month if the prediction was accurate.

JULY

"In the British Isles, Saint Swithun's day (July 15) is said to forecast the weather for the rest of the summer. If St. Swithun's day is dry, then the legend says that the next forty days will also be dry:

"St.Swithin's Day if thou be fair, 'Twill rain for forty days no mair; St. Swithin's Day if thou dost rain, For forty days it will remain. If however it rains, the rain will continue for forty days."

Here is one forecast that does seem to have scientific evidence. In the middle of July, a jet stream settles into a pattern which holds reasonably steady until the end of August. If the Jetstream lies north of the British Isles and high pressure moves in, arctic air and Atlantic weather systems predominate.

In Russia, the weather on "Protecting Veil" day is an indication of the severity of the coming winter.

On St. Paternian's day in Romagna, the dog's tail wags.

This proverb was an indication that the wagging tale means cold weather is coming.

The farmers claim that: As St. Quirinus' Day goes (March 30), so will the summer.

As you have probably noticed many of these proverbs are associated with religious holidays. During this time in history, this would not be unusual as religion felt that the weather was determined by the gods.

ANIMAL SIGNS

COWS

When cows are lying down in a field, rain is on its way.

This may not be a scientific prediction, and yet animal behavior at times seems to predict the weather.

https://en.wikipedia.org/wiki/Weather_lore

A cow with its tail to the West makes the weather best,
A cow with its tail to the East makes the weather least

Cows do not like the wind blowing in their faces so they stand with the rear ends to the wind. With the westerly wind means continuing fair weather, while the easterly wind indicates unfavorable weather. The cow has now become a weather vane or "Cowvane" by predicting the weather.

PETS EATING GRASS

Cats and dogs eat grass before a rain.

This is a false statement as cats and dogs sometimes do eat grass but it has nothing to do with the weather. Cats and dogs are not entirely carnivorous.

PLANT SIGNS

Onion skins very thin
Mild winter coming in;
Onion skins thick and tough
Coming winter cold and rough

Interesting, but no scientific proof.

HOW TO PREDICT THE WEATHER WITHOUT A FORECAST

https://www.wikihow.com/Predict-the-Weather-Without-a-Forecast

There are four methods that are helpful in predicting the weather. They are:

Observing the wind and air.

Observing the clouds.

Observing the Sky.

Observing Animal Behavior and the Community.

If you enjoy the outdoors, the following predictions can be most helpful as they are fairly accurate.

OBSERVING THE WIND AND AIR

The direction of the wind is important. Weather usually moves from the west to the east. Westerly winds indicate good weather. Easterly winds usually mean bad weather is coming.

Tossing a few small blades of grass, or anything light and small, into the air is an easy way to determine direction of the wind.

Some like to wet the finger and hold it out to the wind. The side of the finger that gets cool is the direction of the wind.

Toss the grass is better than wetting the finger with the mouth may not necessarily be sanitary and not as accurate as the grass in the air.

RISING SMOKE

Air pressure determines the direction the smoke will rise. In high pressure, the smoke will go directly up into the air. It is the opposite in low pressure as the smoke often will spiral back down to the fire or to a lesser extent does not rise straight up.

CALM CONDITIONS

Be alert for calm conditions. Before the storm, low-pressure systems can push out the areas normal wind patterns. This creates the calm before the storm. If you are near water, the water will be calm and fairly still. The calm signals a storm coming as may be indicated by dark clouds moving into the area.

SMELL THE AIR

Smell the air with a deep breath. The air smells wet right before a storm. You may also smell a compost effect as plants release their waste. Smell this and a storm is coming.

If you are near a swamp you may smell swamp grass prior to the storm. Swamp gas smells like rotten eggs because of the decaying vegetation

HUMIDITY

High humidity often precedes the storm. Frizzy hair, curling leaves, and swollen wood are signs of an approaching storm.

Pinecones are good indicator of weather. If it is humid, the pinecones are closed. If it is dry, the pinecones are open.

If you live in a high humidity area. also rely on other observations.

OCEAN SWELLS

If you are near the ocean look for ocean swells. These swells are caused by winds blowing a storm system from out over the sea. Rain may be coming.

TYPES OF CLOUDS

Clouds are good weather indicators. High white clouds indicate good weather and naturally dark and low clouds mean rain or storms.

Wispy white clouds mean clear weather.

Flat clouds mean the air is stable, while fluffy clouds mean that the air is unstable.

Small puffy clouds may presently look calm but can build up over the day for the possibility a storm is brewing.

HIGH OR LOW CLOUDS

High clouds means the weather is further away and possibly not a threat until much later. Lower clouds mean bad weather is close. Sometimes as clouds move in lower and lower, the weather becomes more immediate.

THE COLOR OF CLOUDS

Check the color of the clouds

Black clouds mean approaching storm with weak winds.

Brown clouds mean approaching storm with strong winds.

White clouds usually mean good weather but there is a possibility of bad weather later on.

Gray clouds usually mean a light storm over a large area and will settle for a while.

GATHERING CLOUDS

Gathering lowering clouds means bad weather is approaching. If the clouds are rising and spreading it is an indication that the weather is clearing.

RED SKY

Weather moves from west to east. The sun rises in the east and sets in the west, therefore if you see a red sky in the morning it usually means clear weather in the East where the sun is rising, but it is bad weather in the West making the sky look red and the weather may be approaching your area.

The redness in the sky can be from an orange to a deep red.

If you have a red sky in the evening you can rest easy.

This means that there are clear skies in the West, and is moving towards your area while the bad weather is to the east moving away from you.

You can use the following rhyme: *"Red sky at night is a shepherd's delight. Red sky in the morning is a shepherd's warning."*

A RAINBOW

A storm may be moving your way as a rainbow to the west means that the sun's rays, from the East, is striking the moisture in the West. This means the bad weather is moving from the West to your area in the East.

Remember the old saying:

Rainbow in the morning gives you fair warning.

THE MOON

If the moon is easy to see in a clear sky, it could mean that the weather is cooling. If you see a wide halo around the moon there is a possibility of rain.

Ring around the moon? Rain real soon.

A ring around the moon means a warm front is approaching with a good possibility for rain. The ring is caused by ice crystals that are passing over the moon.

A double halo around the moon can signal strong winds and a coming storm.

Clear moon, frost soon.

The clear moon, and clear sky, means that there is no clouds to keep the heat down on the earth. This gives a cooler earth and if it is cold enough we have frost.

STARS

At night, look at the stars. More than 10 stars means that if a storm is approaching it will be light. Fewer than 10 stars the storm could be heavy.

A lack of stars means excessive cloud coverage and weather is moving in.

ANTS

When ants build up their mounds with steep sides storm is approaching.

BIRDS FLYING HIGH OR LOW

When the air pressure falls before a storm, the birds feel discomfort in their ears and as a result they fly closer to the ground or perch on lower tree branches or power lines.

Also, birds feed on ground insects before the approaching storm.

High flying birds indicate fair weather.

Seagulls perched on the beach often indicate a storm is coming.

Watch for large groups of roosting birds.

Singing and chirping birds indicate good weather while silent birds indicate bad weather.

MIGRATING BIRDS

Birds can sense air pressure and will time their migrations to good weather. If you see flocks of birds migrating in the sky, then the weather will likely be good that day.

If you see birds eating during the storm and is a good chance the storm will be long. If the storm is going to be short the birds will usually wait and eat after the storm.

Birds are excellent weather predictors as they sense the pressure patterns.

BEES AND BUTTERFLIES

For their safety, bees and butterflies return to their homes before the storm. If you do not see any bees or butterflies where they are usually located, like in a field of flowers, then you may assume the possibility of a storm coming.

COWS

When cows lie down in the pasture it indicates a rainstorm. Since the weather cools down before a storm, the cows like to lie down. This may be an indication that the soil is warmer than the surrounding air.

This prediction applies only to cows and not other livestock.

SNAKES

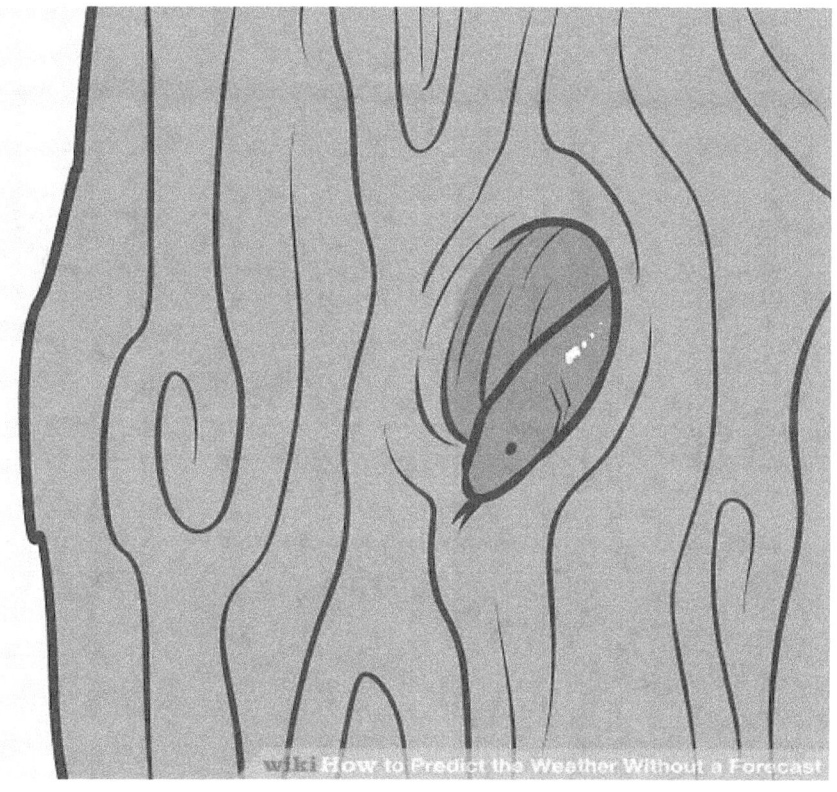

Snakes will leave their nests before bad weather even in the middle of winter. If you see snakes in unexpected places, or when snakes are normally in their nests, it is a sign of bad weather.

Snakes can predict earthquakes. If you see a snake out of its nest behaving erratically, it may be an indication of an earthquake.

Snakes are cold-blooded and usually come out of their nests during sunny warm weather. They do not like the cold. A snake out of its nest when it is not warm or sunny usually means bad weather is ahead.

TURTLES TORTOISES

Tortoises will seek higher ground before a storm. If you are lucky, you may see them on the road one or two days before rain.

WEATHER SAFETY

If you predict or suspect approaching weather, it is best to take safety measures immediately. Delay and it may be to late.

HURRICANE OR TORNADO WITH NO SHELTER

For a tornado get in a ditch, or similar excavation. Go into a fetal position and wait for the storm to pass.

HURRICANE OR TORNADO WITH SHELTER.

Seek a basement or storm cellar. If neither is available go to lowest level of the building and to an inside room with no windows. Place as many walls as possible between you and the outside. Cover yourself with something like pillows, a mattress, or blankets, etc.

HURRICANES

For a hurricane try for high ground for protection from flooding. Be alert for flying debris, falling trees, animals and snakes moving into your area.

EXPECTING THUNDER

There will be heavy dark clouds.

AN APPROACHING TORNADO

Sky will be dark with a possible greenish tint and the air will be still. There may even be wall clouds of debris and possibly hail.

PROTECTION DURING A THUNDERSTORM

The first protection is to seek protection immediately. Do not stand near a tree, avoid power lines and get indoors. Do not use a tree to prevent getting wet.

Seek shelter in a car with a metal roof. Close windows and doors. Do not lean or touch anything metal. Do not use the radio or TV.

Stay away from small structures like public restrooms or rain shelters. Get pets indoors.

Stay away from windows and go to inner rooms of the structure or house.

Do not touch metal or electrical appliances. Do not use landline phones as the wires conduct electricity from the lightning.

Stay out of bathtubs, showers and swimming pools.

Do not be in a hurry to go outside even though the rain has stopped, lightning is still possible.

THUNDERSTORM PROTECTION OUTDOORS

Move to lower elevation so that lightning strikes the higher elevated areas.

Avoid large open spaces where you are the tallest object on the ground. Golf courses can be dangerous as many people have been struck by lightning on golf courses. Metal golf clubs may attract lightning.

If you are fishing or swimming, get out of the water immediately. Water seems to attract lightning.

If you're in a group of people, spread out.

Remove backpacks or any other objects with metal.

Use the lightning crouch by squatting down with your feet together and your head tucked into your chest between your knees to make yourself as small as possible.

Riverboats may be helpful as rubber is a poor electrical conductor.

COOL GUSTS OF WIND

If you feel a sudden gust of wind, it may indicate convection currents of wind by nearby storms.

BUMPY CLOUDS

This is usually indicative of rainstorm.

DOGS GET SCARED BEFORE A STORM

Dogs have supersensitive hearing and they can hear distance thunder that humans do not hear.

HIGH PRESSURE

Usually means clear sky, dry air and sunshine.

ANIMAL WEATHER PREDICTORS

Animals can often sense the weather that we have noticed in some of the earlier writings but they are not really able to predict the weather.

CALCULATE THE DISTANCE FROM LIGHTNING

This is a good safety procedure.

Watch for the flash of lightning. Count the number of seconds until you hear thunder. If you don't have a watch use the old second count by of one 1000, two 1000, etc.

Sound travels 1 mile every five seconds and 1 km every three seconds to determine distance from lightning divide the number of seconds by five and your answer is in miles. Divide by three if you want the answer in kilometers.

For example:

If you counted 18 seconds divide 18 x 5 unit 3.6 miles. For the distance of lightning in kilometers divide 18 x 3 you get 6 km.

METEOROLOGICAL SIGNS

Rain before seven, clear by eleven.

Not necessarily so.

The following is from:

http://www.tww.id.au/weather/contrib.html#march

Observing the clouds.

A red sky at night is a sailor's delight. A red sky in the morning is a sailor man's warning.

Red sky at night, shepherds delight. Red sky in the morning, shepherds warning.

The evening red and morning gray are sure signs of a fine day, but the evening gray and the morning red, makes the sailor shake his head.

Evening red and morning gray will set the traveler on his way;
but evening gray and morning red will bring down rain upon his head.

Evening red and morning gray. Two sure signs of one fine day.

Evening grey and morning red, put on a hat or you'll wet your head.

A setting red sun means it'll be hot tomorrow.

If the sun goes pale to bed, t'will rain tomorrow, it is said.

If red the sun begin his race, be sure the rain will fall apace.

If red, the sun set in gray the next will be a rainy day.

If the morning sky is red, the ewe and her lamb will go wet to bed.   A child's interpretation.

If clouds are gathering thick and fast, keep sharp look out for sail and mast, But if they slowly onward crawl, Shoot your lines, nets and trawl.

When clouds are gathering thick and fast, keep sharp lookout for sail and mast; but if they slowly onward crawl, out with the dories, nets, or trawl.

The higher the clouds the better the weather.

High clouds indicate fine weather will prevail. Lower clouds mean rain.

Mackerel sky, mackerel sky - never long wet, never long

dry.

Herringbone sky, neither too wet nor too dry (clouds have a herringbone pattern).

Herringbone sky, won't keep the earth 24 hours dry.

Mare's tails (clouds); storms and gales.

Clouds known as mares' tails by sailors of old, are fallstreaks. They are fibrous, hooked-shaped clouds that are composed of ice. Although not uncommon, the fallstreak is somewhat of an intriguing mystery.

Mackerel sky; not 24 hours dry.

Horses' manes and mares' tails-- sailors soon shall shorten sails.

Mackerel sky and mares' tails make lofty ships carry low sails.

Mackerel scales and mare's tails make lofty ships carry low sails.

Mackerel skies and mare's tails, make ships carry short sails.

Backing winds and mares tails make tall ships carry small sails!

If clouds look as if scratched by a hen. Get ready to reef your topsails then.

When clouds appear like rocks and towers, the earth's refreshed with frequent showers.

When clouds look like rocks and towers, the earth will be refreshed by showers.

If wooly fleeces deck the heavenly way, be sure no rain will mar a summer's day .

_____

_____

OBSERVING THE WIND

When the wind shifts against the sun, trust it not, for back it will run.

When the wind is from the south the rain's in its mouth.

When rain comes before the wind, halyards, sheets and braces mind, but when wind comes before rain, soon you may make sail again.

When rain comes before the wind, Dories, gear and vessel mind; when wind comes before the rain, soon you'll make the set again.

When the wind is blowing in the North, no fisherman should set forth,

When the wind is blowing in the East, 'tis not fit for man nor beast,

When the wind is blowing in the South, it brings the food over the fish's mouth,

When the wind is blowing in the West, that is when the fishing is best!

Wind from the East fish bite least, wind from the West fish bite best.

With rain before wind, stays and topsails you must mind, but with the wind before the rain, your topsails you may set again.

If the wind is in the east, they bite the least.

If the wind is in the west, they bite the best.

If wind in the south, throw bait in their mouth.

If wind in the north, stay home by the hearth.

Wind before the rain, turn and shoot again. Rain then the wind, pick up and go in.

Every wind has its weather.

The wind in the west suits everyone best.

No weather is ill, if the wind be still.

A little rain, stills a great wind.

OBSERVING RAIN, DEW, SNOW, FOG AND RAINBOWS

If it's foggy in the morning then it'll be a sunny day.

When the fog goes up the hill it takes the water from the mill. When the fog comes down the hill, it brings the water to the mill.

Rain before seven, clear by eleven.

When dew is on the grass, no rain will come to pass.

A little rain, stills a great wind.

Snow like meal, snow a great deal.

Wind before the rain, turn and shoot again. Rain then the wind, pick up and go in.

When rain comes before the wind. Halyards, sheets and braces mind, But when wind comes before rain, soon you may make sail again.

When rain comes before the wind, dories, gear and vessel mind; when wind comes before the rain, soon you'll make the set again.

If there is dew upon the ground, no rain that day will be found.

Unless it rained the night before and then you had better keep score.

Rainbow in the morning, travelers take warning; rainbow at night, traveler's delight.

Rainbow at night, sailors delight; rainbow in the morning, sailors warning.

Rainbow in the eastern sky, the morrow will be dry. rainbow in the west that gleams, rain falls in streams.

OBSERVING PLANTS, BIRDS, ANIMALS AND INSECTS

Ash before oak in for a soak, Oak before ash in for a splash.

Seagull, seagull sit on the sand. It's never good weather when you're on the land.

Seabirds, stay out from the land, we won't have good weather while you're on the sand.

An Australian rain prediction that works in winter, "When the kookaburra's call, the rain will fall".

When the trees begin their dance. Of rain, there is a great big chance.

When the ass begins to bray, be sure we shall have rain that day.

When leaves show their back, rain we won't lack.

Moss dry, sunny sky, moss wet, rain we will get.

Bees do not swarm before a storm.

When the rooster goes crowing to bed, he will rise with a watery head.

OBSERVATIONS AT NIGHT

A change in the moon brings on a change in the weather.

If all stars are out at night, it will be a nice day tomorrow.

When halo rings the moon or sun, rain's approaching on the run.

If a circle forms around the moon, 'twill rain soon.

If the moon holds water it will be dry. If water from it can leak rain is nigh.

OBSERVING THE BAROMETER

Barometer is a scientific instrument used in meteorology to measure atmospheric pressure. Pressure tendency can forecast short-term changes in the weather.

In 1643/5, the barometer was invented as a 3 foot glass tube. In 1884, the aneroid barometer was invented and replaced the 3 foot glass tube.

This is the original barometer by Johann Wolfgang Von Goethe,

This device was known as "weather glass" or "Goethe barometer".

Source: Barometer - *https://en.wikipedia.org*

"The weather ball barometer consists of a glass container with a sealed body, half filled with water. A narrow spout connects to the body below the water level and rises above the water level. The narrow spout is open to the atmosphere. When the air pressure is lower than it was at the time the body was sealed, the water level in the spout will rise above the water level in the body; when the air pressure is higher, the water level in the spout will drop below the water level in the body. A variation of this type of barometer can be easily made at home." (as shown at end of book).

MERCURY BAROMETERS

A mercury barometer has a vertical glass tube closed at the top sitting in an open mercury-filled basin at the bottom. The weight of the mercury creates a vacuum in the top of the tube known as Torricellian vacuum. Mercury in the tube adjusts until the weight of the mercury column balances the atmospheric force exerted on the reservoir. High atmospheric pressure places more force on the reservoir, forcing mercury higher in the column. Low pressure allows the mercury to drop to a lower level in the column by lowering the force placed on the reservoir. Since higher temperature levels around the instrument will reduce the density of the mercury, the scale for reading the height of the mercury is adjusted to compensate for this effect. The tube has to be at least as long as the amount dipping in the mercury + head space + the maximum length of the column.

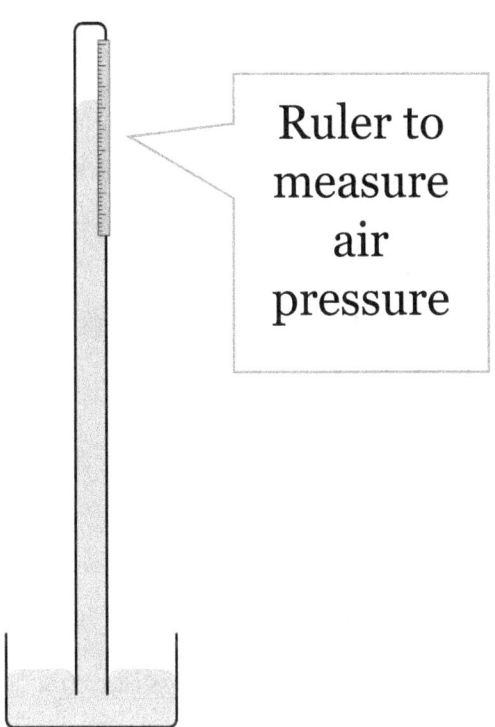

Ruler to measure air pressure

Source: File:MercuryBarometer.svg - https://en.wikipedia.org

ANEROID BAROMETERS

This is the modern barometer for measuring air pressure. It is compact and readily transportable.

Source: File:Modern Aneroid Barometer.jpg - https://en.wikipedia.org

PREDICTIONS BY THE BAROMETER.

When the glass falls low, stand by for a blow;
When it slowly rises high, all the light sails you may fly.

When the glass falls low, prepare for a blow;
When it slowly rises high, lofty canvas you may fly.

Long foretold, long last, short notice, soon past,
Quick rise after low, sure sign of stronger blow.

Quick rise after low, indicates a stronger blow;
Long foretold, long last, short notice, soon past.

Fast rise after low,
Foretells a stronger blow.

At sea with low and falling glass soundly sleeps a careless ass, only when it's high and rising, truly rests a careful wise one.

OTHER OBSERVATIONS

When Windows won't open, and the salt clogs the shaker, The weather will favor the umbrella maker!

When you hear the train, it will rain. This is an interesting one but is logical. Sound transport better with high humidity. People who lived about 3 miles from a railroad track could hear the train whistle on high humidity days but not on low humidity days. Naturally, the high humidity indicated rain.

Thunder before seven, rain before eleven.

OBSERVATIONS FOR THE SEASONS

The following seasonal observations apply only to the northern hemisphere as the southern hemisphere has seasons opposite the northern hemisphere.

When March blows it's horn, your barn will be filled with hay and corn. It means that a thunderstorm in March will forecast to harvest. By July 4, the corn is knee-high and you will have a good crop.

January brings the snow, makes our feet and fingers glow.

February brings the rain, thaws the frozen lakes again.

March brings breezes sharp and chill, shakes the dancing daffodil.

April brings the primrose sweet, scatters daisies at our feet.

May brings flocks of pretty lambs, sporting round their fleecy dams.

June brings tulips, lilies, roses, fills the children's hands with posies.

Hot July brings thunder-showers, apricots, and gilly-flowers.

August brings the sheaves of corn; then the harvest home is borne.

Warm September brings the fruit; sportsmen then begin to shoot.

Brown October brings the pheasant, then to gather nuts is pleasant.

Dull November brings the blast-- hark! The leaves are whirling fast.

Cold December brings the sleet, blazing fire, and Christmas treat.

Ice in November to bear a duck, the rest of the winter will be slush and muck.

April showers bring May flowers. March winds bring April showers.

March comes in like a lamb, goes out like a lion. March comes in like a lion, goes out like a lamb.

If there's no dew on the grass in the morning, rain will be here within 24 hours. This seems merited in Southern California especially during the rainy season.

St. Swithin's day, if thou dost rain, for forty days it will remain; St. Swithin's day, if thou be fair, for forty days 'twill rain na mair.

FOLKLORE WEATHER BY PEOPLE

http://www.tww.id.au/weather/contrib.html#march

This part is about observations and findings by various people who are not scientists but out of curiosity developed their own predictions.

http://www.tww.id.au/weather/forecast.html

FAIRWEATHER LEWIS

The Ruling Days.

January 6, 2011 by Faire.

In the knobs, tradition says that the twelve days beginning on December 26 and ending on January 6 are the "ruling days". On those days, the old people say, the weather in the coming year depends, each day corresponding to a month. Thus, December 26.

A rainbow in the morning is the Shepherd's warning.

A rainbow in the night is the Shepherd's delight.

2. A wet and windy May fills the barn with corn and hay.

3. A mackeral sky is never dry.

4. If March comes in like a lion it goes out like a lamb.

5. Soft April showers and bright May flowers will bring the summer back again.

6. When the cuckoo calls from a leafless thorn sell your cow and buy corn.

7. A green Christmas makes a fat churchyard.

8. A wet May and a fine June, makes the farmer whistle a merry tune.

9. If the ash should bud before the oak,
Then we're sure to have a soak,

If the oak should bud before the ash,
Then we'll only have a splash.

Old weather proverbs and sayings to forecast the weather accurately was important for farm families. The people just used observations to develop their own forecasts.

Here's what a 1913 article in a local newspaper about weather proverbs:

OLD PROVERBS ON WEATHER ARE TRUE

Ancient Sayings Based on Experience Are Approved by Uncle Sam's Scientific Investigators.

The idea that old weather proverbs and traditional natural signs are of no value in these days of scientific weather forecasting is not sustained by such an eminent authority as W.J. Humphreys, professor of meteorological physics in the United States Bureau.

He gives credit to the weather prescience of farmers, fishermen, woodsmen and others who make a practice of depending on natural signs to give them knowledge of impending weather changes.

Quoting the jingle about a sailor's warning and a sailor's delight, Professor Humphreys says:

"If the evening sky, not far up but near the western horizon, is yellow, greenish, or some other sort wave-length color, then all the greater the chance for clear weather, for these colors indicate ever less condensation and therefore drier air than does red."

Professor Humphreys says a good word for such old-time proverbs as:

Frost year, fruit year.

Year of snow, fruit will grow.

A year of snow, a year of plenty.

"That these and similar statements commonly are true," he writes, "is evident from the fact that a more or less continuous covering of snow, incident to a cold winter, not only delays the blossoming of fruit trees until after the probable season of killing frosts but also prevents that alternate thawing and freezing so ruinous in fruit. In short, as another proverb puts it, *a late spring never deceives.*

The appearance of the moon depends upon the conditions of the atmosphere.

Clear moon, frost soon, and *moonlit night has the heaviest frosts* and others of this class are true enough he says, because on the clearest nights the cooling of the earth's surface by radiation is greatest, and hence most likely to cause, through the low temperature reached, precipitation in the form of dew or frost.

Milton Evening Standard (June 21, 1913)

WEATHER PROVERBS

"Halo around the sun or moon, rain or snow soon."

"A rainbow afternoon, good weather coming soon."

"Horses run fast before a violent storm or before windy conditions." Weather Proverb

"When the stars begin to huddle, the earth will soon become a puddle." Weather Proverb

"Thunder in the morning, all day storming. Thunder at night is the travelers delight."

"No weather's ill, if the wind be still."

"Clear moon, frost soon."

"Red Sky at night, sailor's delight. Red sky in the morning, sailor take warning."

The moon and the weather may change together, but a change of the moon, will not change the weather."

"The higher the clouds the better the weather."

"A tough apple skin means a hard winter."

"Moon rising clear, fair weather is near." Weather Proverb

"When the wind blows from the west, fish bite best. When it blows from the east, fish bite least."

"A year of snow, a year of plenty."

"Cold is the night when the stars shine bright."

"Look for rain when the crow flies low.

"If March comes in like a lion, it will go out like a lamb."

"If bees stay at home, rain will soon come, if they fly away, fine will be the day."

"Smoke curling downward, poor weather.

"The more cloud types present, the greater the chance of rain or snow."

"The squeak of the snow will the temperature show."

"One crow flying alone is a sign of foul weather; but if crows fly in pairs, expect fine weather."

"Rainbow in the morning gives you fair warning."

"Fair weather cometh out of the north."

"Chimney smoke descends, our nice weather ends.

"When the sun shines while raining, it will rain the same time again tomorrow."

"A wet January, a wet spring."

"No matter how long the winter, spring is sure to follow."

"When your joints all start to ache, rainy weather is at stake."

"Fair on September first, fair for the month."

"Mare's tails and mackerel scales make tall ships take in their sails."

"The sudden storm lasts not three hours, the sharper the blast, the sooner 'tis past."

"Expect rain and maybe severe weather when dogs eat grass."

"A ring around the sun or moon means rain or snow coming soon."

"April showers bring May flowers."

"Fish bite best before a rain."

"If shooting stars fall in the south in winter, there will be a thaw."

"If autumn leaves are slow to fall, prepare for a cold winter."

"If you see the underside of the leaves in the gentle breeze, it will rain before your sneeze."

"If wasps build their nests high, the winter will be long and harsh." Weather Proverb

"Seagull, seagull, sit on the sand, it's a sign of rain when you are at hand."

MORE

WEATHER PROVERBS

http://boatsafe.com/nauticalknowhow/weather_proverbs.htm

We have been attempting to forecast the weather since the beginning of recorded history. Long before the invention of radar and other meteorological tools, people relied upon "natural" clues to approaching weather. Many of these have a scientific basis and it can be explained why they "work," others have no such basis but often prove to be true.

Perhaps the most often quoted weather proverb among mariners is:

Red sky in morning, Sailors take warning.

Red sky at night, Sailors' delight.

A red sky at night (when the sun is to the west) is caused by light passing through dust particles in the air to the west. Dust indicates dry weather and since most weather changes come from the west, a red sky at night usually indicates dry weather approaching.

A red sky in the morning, however, indicates that the dry air has moved away. A gray sky at night means that the western air is filled with moisture and it will likely rain soon.

The first recorded use of this system of weather forecasting can be found in the Bible. In Matthew 16.2-3, Jesus says to the fishermen, "when it is evening, you say, 'It will be fair weather, for the sky is red.' and in the morning 'It will be stormy today, for the sky is red and threatening." Since it has lasted so long, we think there must be something to it.

Other variations on this theme include:

Evening red and morning gray, help the traveler on his way.
Evening gray and morning red bring down a rain upon his head.

Rainbow in the morning gives you fair warning.

The sun is in the east in the morning, the shower and associated rainbow are in the west. Since weather generally moves from west to east, rain is approaching.

Beware the bolts from north or west;
In south or east the bolts be best.

Same reasoning as the above.

Rainbow to windward, foul fall the day;
Rainbow to leeward, rain runs away.

If the wind is coming from the direction of the rainbow, the rain is heading toward you. Conversely, if the rainbow is in the opposite direction, it has passed you.

Mackerel skies and mares' tails
Make tall ships take in their sails.

Cirrus clouds (mackerel skies or clouds that looked as if they'd been scratched by a hen, according to the old-timers) often precede a warm front which brings winds and rain.

When halo rings the moon or sun
Rain's approaching on the run.

The halo is caused by high cirrostratus (ice crystal) clouds that are indicative of an approaching warm front and predict rain within 20-24 hours. The U.S. Weather Service confirms that rain follows about 75 percent of sun halos and about 65 percent of moon halos.

The higher the clouds
the better the weather

These clouds generally indicate both dry air and high atmospheric pressure - usually associated with fair weather. Lowering ceilings indicate rain.

A wind from the south
has rain in its mouth

A south wind blows in advance of a cold front and also blows over the east quadrant of an approaching low pressure cell.

Seagull, seagull, sit on the sand,
It's a sign of rain when you are at hand

In general, birds roost more during a period of low pressure. Before a hurricane, flocks of birds will be seen roosting. Take off may be harder when the pressure is low or the air is thinner because the natural updrafts are lessened.

Some weather proverbs published in 1883 by the War Department (no explanation given):

Buzzards flying high indicate fair weather.

One crow flying alone is a sign of foul weather; but if crows fly in pairs, expect fine weather.

When porpoises and whales spout about ships at sea, storm may be expected.

Two full moons in a calendar month bring on a flood.

Comets bring cold weather. If shooting stars fall in the south in winter, there will be a thaw.

Lightning under the North Star will bring rain in three days.

Interesting one that warrants further investigation:

When the bubbles of coffee collect in the center of the cup, expect fair weather. When they adhere to the cup, forming a ring, expect rain. If the bubbles separate without assuming any fixed position, expect changing weather.

Thanks to Norman Westrick for addressing a possible answer to this often quoted, but perhaps less understood, weather proverb.

Norman's theory:

As I'm sure you know, liquids have surface tension. They also tend to adhere to objects. That is why you can fill a glass to just a little over the top without it spilling.

When coffee is hot, it creates a small amount of pressure on the under side of the surface. If the barometric pressure is low, the pressure in the cup (created by the heat) will overcome the atmospheric pressure and cause the surface of the coffee to be convex and the bubbles will settle to the edge of the cup. Low barometric pressure indicates weather deteriorating.

If the barometric pressure is high, the pressure in the cup will be depressed by the atmospheric pressure and the surface of the coffee will be concave and the bubbles will settle to the center of the cup. High barometric pressure indicates clear weather.

The last part of the proverb (bubbles separate without assuming any fixed position) would be when the barometer is on its way up or down and has found that happy middle ground where the pressure above and below the surface of the coffee is about the same.

Seems logical.

WEATHER URBAN LEGENDS AND FOLKLORE

http://www.ashevillelist.com/weather_folk_sayings.htm

Folklore for predicting the weather are urban legends, old wives tales, superstitions, folk tales and old sayings. Most folklore has its foundations in observation. Long ago weather lore became an important part of religion. Priests decided what dates to sow seed or harvest. It was important to interpret weather signs. Weather lore ranges from common sense to superstitions based upon observation and more often wishful thinking. Many folklore tales contradict one another. They have never been proven to be true. However, mountain folk in the Asheville area (North Carolina) who swear they are true.

Dogwood Winter - Every year when the Dogwood trees bloom, it gets real cold. (Never seems to fail, it always gets cold in the NC mountains when the dogwood bloom)

Blackberry Winter - Every year when the blackberries bloom, it gets real cold. (Never seems to fail, it always gets cold in the NC mountains when the blackberry bloom).

Never plant crops on a Sunday, they will fail if you do.

Plant crops on Good Friday for a good harvest.

If yellow jackets are building their nests above ground, then it will be a wet winter.

If the wooly worm has a lot of wool, it will be a bad winter.

If the squirrels and birds are feeding in the winter, expect a bad snow storm

For every foggy morning in August, it will snow that many days that following winter.

If it rains when the sun is out, the devil is beating his wife

If it lightning in winter, within 10 days it will snow.

If the first thunder is from the east, winter is over.

Rain and wind increase after a thunderclap.

A ring around the sun or moon is a sign that rain is approaching.

Do business with men when the wind is in the northwest

People who have chronic ailments such as arthritis, gout, rheumatism and corns can predict a change in weather by the increase of their aches and pains.

Red Sky

Red sky at night, sailors delight. Red sky in the morning, sailors take warning

Red sky at night, sheppard's delight, Red sky in morning Sheppard's take warning

From the bible, Matthew 16:2-3 - When it is evening, ye say, It will be fair weather: for the sky is red. And in the morning, it will be foul weather: for the sky is red.

RED SKY IN THE MORNING, SAILORS TAKE WARNING... AND OTHER NAUTICAL LORE SAYINGS

January 6, 2014 by Basil Karatzas

http://gcaptain.com/nautical-lore-red-sky-morning-warning/

Some of the poems and rhymes are repeated from earlier readings but the explanation and the reason in maybe a little different and maybe more scientific so some of them are repeated again for better understanding.

(c) Shutterstock/apiguide

Ever since people tried to navigate the seaways, the prevailing weather has always been very crucial to the success of expeditions in steering away from trouble. Tall ships were completely dependent on wind so understanding the weather elements were important to any good navigator's skill set.

Early records to understand the weather, from Aristotle's *Meteorologica* around 340 B.C., a philosophical treatise with theories about the formation of rain, clouds, hail, wind, thunder, lightning, and hurricanes, meteorology has evolved into a highly quantitative science.

Think of the Earth as a huge sphere of 12,800 kilometers (8,000 miles) in diameter covered with a 40-kilometer skin of various gases, whose concentration varies both spatially and temporarily. This sphere has a bumpy surface and is rotating all the time. This often determines the movement of wind and air. The earth's tilt gives us seasons. The earth is heated by the sun, 93 million miles away, with about eight million quadrillion BTUs of solar energy each year. Interaction of these factors, the sun, wind, and earth's surface, cause the weather.

Long before the National Oceanic and Atmospheric Administration (NOAA) in the US and the Met Office with the Shipping Forecast on BBC's Radio 4 in the UK, sailors and navigators observed weather patterns to become weather forecasters.

At sunrise and sunset, sunlight is set low on the horizon and travels disproportionally longer distance in the lower atmospheric strata in order to reach the earth surface; thus, sunlight spending more time traveling the atmospheric strata determine the weather, and thus providing more clues for its forecast.

At noon, sunlight hits the surface of the Earth vertically – at least at the Equator – penetrating uniformly all atmospheric strata, and providing less of a clue about weather changes. Preponderance of dry dust particles in the air is a proxy of lack of water vapors in the air (predecessor to rain), and thus dry air particles act as a prognosticator for lack of immediate raining.

In the Northern Hemisphere and around the mid-latitudes ('Horse Latitudes', where becalmed vessels often threw overboard horses due to lack of water onboard the vessels).

Usually prevailing winds move from west to the east; easterly moving dry dust particles were located westerly or easterly to the observer, it has been a fair quick rule of thumb about weather forecasting.

Based on these rudimentary principles, here are few interesting nautical weather sayings for old salts and landlubbers alike:

Red sky at night, sailor's delight,

Red sky in the morning, sailors take warning.

Variations of the theme:

Evening red and morning gray, help the traveler on his way.

Evening gray and morning red bring down a rain upon his head.

Orange or yellow, can hurt a fellow.

These are perennial favorites of weather sayings with interesting scientific explanation behind it: in the mid latitudes of the Northern Hemisphere, typically weather moves from west to east, blown by the westerly trade winds, meaning that storm systems generally move in from the West.

The colors we see in the sky are due to the rays of sunlight being split into colors of the spectrum as they pass through the atmosphere and ricochet off particles and water vapor in the atmosphere. The amount of dust particles and water vapor in the atmosphere are good indicators of weather conditions. They also determine which colors we will see in the sky.

During sunrise and sunset, the sun is low in the sky and it transmits light through the thickest part of the atmosphere. A red sky suggests that lower atmospheric strata are loaded mainly with dust particles (when atmospheric pressure is high, the lower air holds more dust than water vapors) and low in water vapor concentration. When we see a red sky at sunset, this means that the setting sun is sending its light through a high concentration of dust particles. This usually indicates high pressure and stable air coming in from the west, meaning that effectively good weather will follow.

We see the red, because red wavelengths (the longest in the color spectrum) are breaking through the atmosphere. The shorter wavelengths, such as blue, are scattered and broken up.

A red sunrise reflects the dust particles of a system that has just passed from the west. This indicates that a storm system may be moving to the east. If the morning sky is a deep fiery red, it means a high water content in the atmosphere. So, rain is on its way.

A red sky in the morning can be caused by the dawn light bouncing off cirrus ice crystals in the upper atmosphere. Cirrus clouds can be at the leading edge of a frontal system and so this can also work to signal poor incoming weather.

This weather saying has been referred to in the Bible (Matthew XVI: 2-3,) when Jesus said to the fishermen, "When in evening, ye say, it will be fair weather: For the sky is red. And in the morning, it will be foul weather today; for the sky is red and lowering."

Several centuries later, Shakespeare in his play *Venus and Adonis* says: *'Like a red morn that ever yet betokened, Wreck to the seaman, tempest to the field, Sorrow to the shepherds, woe unto the birds, Gusts and foul flaws to herdmen and to herds.'*

A comparable saying, incorporating the gray color has it as:

The evening red and morning gray are sure signs of a fine day, but the evening gray and the morning red, makes the sailor shake his head.

Gray sky at night means that the western air is filled with moisture and it will likely rain soon.

Based on similar scientific analysis, also:

A rainbow in the morning gives you fair warning.

In the morning, in the northern hemisphere a rainbow out to the west is caused by the sun in the east refracting on water droplets to the west, similarly to producing red skies. And moisture in the air will be heading east likely to produce rain.

Likewise,

Rainbow to windward, foul fall the day; rainbow to leeward, rain runs away.

A rainbow from where the wind is blowing (from the west, usually, in the Northern Hemisphere) is indicating that water vapors are closing in as they are pushed with the wind, whereas a rainbow to the direction the wind is blowing (leeward), it means that water droplets have already passed the weather observer.

Related to rainbows, there is the saying of:

When the sun shines while raining, it will rain the same time again tomorrow.

The incident of raining while the sun is shining is called 'sun shower' or 'sunshower' and often accompanied by the formation of a rainbow. As per rainbow explanations above, rain from westerly winds is still in the cards.

Mackerel sky, not 24 hours dry,

Or, its variation:

Mackerel sky, mackerel sky, never long wet, never long dry.

Or, another variation:

Mackerel sky (or scales) and mares' tails, make lofty (or tall) ships carry low sails (or, Make tall ships take in their sails).

Mackerel or fish scale cloud formations are high, thin cirrocumulus clouds formed by shifting wind directions and high speeds and are typical of an advancing low pressure system or an approaching storm system or front.

'Mare's tails' is a term used to describe those high cirrus clouds that are caused by strong winds high in the air.

So, it stands to reason that if you have a Mackerel sky and mares' tails together, it is going to be wet and windy.

When a halo rings the moon or sun, Rain's approaching on the run (or, The rain will come upon the run).

or, its variation:

If there is a halo round the sun or moon, then we can all expect rain quite soon.

or,

A ring around the sun or moon, means that rain will come real soon.

or

Halo around the sun or moon, rain or snow soon.

A ring or halo around a bright object like the sun or the moon is caused by refraction of the light through the ice crystals of high cirrus (cirrostratus) clouds. The presence of these ice crystal clouds is often a sign that a weather front is on its way probably bringing rain and the brighter the circle, the greater the possibility.

Cirrus can be the first cloud to appear ahead of a front. The U.S. Weather Service confirms that rain follows about 75 percent of sun halos and about 65 percent of moon halos.

Sun sets Friday clear as bell, Rain on Monday sure as hell.

Unknown explanation.

If clouds are gathering thick and fast,

Keep sharp look out for sail and mast,

But if they slowly onward crawl,

Shoot your lines, nets and trawl.

When the wind is blowing in the North

No fisherman should set forth,

When the wind is blowing in the East,

'Tis not fit for man nor beast,

When the wind is blowing in the South

It brings the food over the fish's mouth,

When the wind is blowing in the West,

That is when the fishing's best!

With the approach of a low pressure front, easterly winds typically pick up, uncomfortably warm, dry, and dusty in the summer and bitterly cold in the winter.

Northerly winds, which follow around a low, are cold and blustery. Sailing in conditions of northerly winds requires expertise and a sturdy vessel capable of handling heavy seas.

Southerly winds typically bring warmer temperatures, and though they may not feed the fish, they do provide pleasant fishing weather. The best is to have a westerly wind blowing since it is likely to persist for some time, the weather should remain fair and clear, and the wind should be relatively constant.

Beware the bolts from north or west; in south or east the bolts be best.

Meaning that storms to port going North that the storm is coming your way (from the west), while storms to starboard have passed.

A wind from the south has rain in its mouth.

On occasion attributed to Benjamin Franklin, a wind from the south usually brings rain and precedes a cold front.

When rain comes before the wind, halyards, sheets and braces mind,

But when wind comes before rain, soon you may make sail again.

or similarly,

With the rain before the wind, stays and topsails you must mind,

But with the wind before the rain, your topsails you may set again.

Winds occur when two masses of air of different pressure come into contact; for westerly prevailing winds in the Northern Hemisphere, winds before the rain indicated that the two masses of air are already in contact and thus the strong winds, with the rain following from the westerly winds.

However, rain, prior to the winds, is indicative of westerly winds blowing to the east without yet reaching the weather front and creating a storm.

No weather's ill, if the wind be still.

Typically, strong winds occur near weather fronts (frontal boundaries) where two masses of air of different pressures come into contact. Winds tend to be stronger near these frontal boundaries. When the wind is still, it tends to be toward the center of high pressure or the center of an air mass, and thus no 'weather illness'. Calm conditions, especially with clear skies, indicate a high pressure area and lack of any phenomena typically associated with weather, such as clouds, wind, and precipitation. However, calm conditions may also result from a circumstance known as 'the calm before the storm', when a large storm cell to the west is sucking up the surface wind in its updraft before it arrives. This situation is readily apparent by looking to the west for the approaching storm.

Similarly,

If clouds are gathering thick and fast,

Keep sharp look out for sail and mast,

But if they slowly onward crawl,

Shoot your lines, nets and trawl.

The sudden storm lasts not three hours

The sharper the blast, the sooner 'tis past.

When weather fronts masses of air of different pressure and temperature forcefully interact, usually such forceful interaction lasts only for a few hours since a new approximate equilibrium of barometric pressures is achieved and thus mitigating the forcefulness of the 'blast'.

Clear moon, frost soon.

or, similarly,

Cold is the night when the stars shine bright.

When there are no clouds to obscure the moon, there are no clouds to 'blanket' the earth's surface and retain any heat that the earth absorbed during the day so, the surface will cool rapidly on a clear night.

If the new moon holds the old moon in her lap, expect fair weather.

When the new moon can be seen along with the outline of the rest of the moon ('the old moon') as in a shadow, then the air must be clear and stable enough for us to see faint objects in the sky. Thus, it means that the weather is fair and is likely to stay that way for a while.

The higher the clouds, the better the weather

For cumulus clouds, nice little woolpacks, with high bases – around 4000 feet or more, rain is unlikely. Similarly, for mackerel appearance clouds (cirrocumulus clouds) with no mare's tails, then again, the weather looks fair. These clouds generally indicate both dry air and high atmospheric pressure – usually associated with fair weather. Lowering ceilings indicate rain. However, ahead of a warm front, high cirrus clouds will be spreading high across the sky, a fore-runner of rain some hours later.

A piece of seaweed hung up will become damp before it rains.

As seaweed naturally absorbs water, as atmospheric humidity increases before it rains, dry seaweed gets damper easier and faster, and thus the 'science' behind the saying.

Similarly,

When ropes twist, forget your haying.

Natural hemp ropes and rigging have a tendency to twist as humidity rises as they get damper with water vapors from the atmosphere, thus indicating that rain will follow soon.

Similarly in terms of explanation,

When the chairs squeak, it's of rain they speak.

Animals, birds and fish have been known to have tried their luck with forecasting the weather:

Seagull, seagull, sit on the sand, it's a sign of rain when you are at hand (or, it's never good weather when you're on the land).

or,

When sea-gulls fly to land, a storm is at hand.

Anecdotal evidence suggests that animals can sense miniscule changes in the environment and their actions accordingly can predict the weather. It's not always clear how animals exactly sense or interpret changes of atmospheric pressure or impending storms, etc, and how their sensory reception can come to their survival instincts. In general, birds and animals roost/being nothalgic more during periods of low atmospheric pressure. Before a hurricane, flocks of birds will be seen roosting; taking off may be harder when the pressure is low or the air is thinner because the natural updrafts are lessened.

Sharks go out to sea at the approach of a wave of cold weather.

Several studies have shown that sharks are known to move to deeper waters before hurricanes and storms, seeking better protection from strong waves and jeopardizing getting launched on land.

When porpoises sport and play, there will be a storm.

Porpoises are aquatic mammals similar in appearance to dolphins; porpoises are not successfully kept in captivity like dolphins. Not much explanation is available why these animals, and also land animals and pets, get more playful before severe weather.

A backing wind says storms are nigh, but a veering wind will clear the sky.

A backing wind is a wind that turns counter-clockwise with height, while a veering wind is a wind that turns clockwise with height.

A backing wind is associated with cold air advection and dynamic sinking (CCBC or CounterClockwise, Backing, Cold air advection). Ahead of a warm front, the wind will back from W or NW to SW, S and even SE. So, this can be a good predictor although not all backing winds will presage a warm front. As a cold front passes, winds will veer from a SW'ly point ahead of the front to NW behind. So, in this case, the wind veer (CVW or Clockwise, Veering, Warm air advection) will be as or after the front has passed. Winds back behind cold fronts.

If wooly fleeces deck the heavenly way, be sure no rain will mar a summer's day.

Fleecy, wool-like white clouds are only a few hundred feet thick, indicating that they have barely developed due to condensation. As such, they are a sign the atmosphere is still relatively stable.

When boat horns sound hollow,

Rain will surely follow.

Sound travelling far and wide,

a stormy day betide.

Sound travels faster in the water than in the air, and a little bit faster in humid rather than in dry air.

Also, sound is absorbed to the greater extent when traveling through humid air rather than dry air. Thus, sounds in humid days – days proceeding rain, have a hollow (echo-y) effect and travel faster.

When the stars begin to huddle, the earth will soon become a puddle.

As water vapors and humidity increase in the air, in advance of rainy weather, smaller stars on the sky cease to be visible, while bigger, brighter ones overwhelm the sky and shine with a blur or a corona around them ('fogginess'), giving the impression of cluster of stars rather individual stars.

When the bubbles of coffee collect in the center of the cup, expect fair weather. When they adhere to the cup, forming a ring, expect rain. If the bubbles separate without assuming any fixed position, expect changing weather.

This was explained earlier.

FOREIGN WEATHER PROVERBS

http://www.bartleby.com/346/16.html

We have more Proverbs but the interesting part is they give the location and country of the proverb. Some of these have been repeated but the least you realize where the adage or proverb originated.

A cloudy sky on Friday and Saturday, says Bhadarri, is a sure precursor of rain. (Behar).

"Bhaddar," Mr. John Christian tells us, "was a local

poet of some fame. He interpreted the signs of the seasons in rhymes which have passed into proverbs.... When very young, he was stolen from his home in Shāhabad by a famous magician or astrologer, who carried him away to his country and adopted him. Bhaddar became so thoroughly proficient in astrology and all the mystic arts, that his patron gave him his daughter in marriage."

A fine Saturday, a fine Sunday; a fine Sunday, a fine week. (English).

"Fine on Friday, fine on Sunday; wet on Friday, wet on Sunday." (French).

"There is never a Saturday without some sunshine." (English).

A foul morn may turn to a fine day. (English).

"If it rains before seven, 'twill cease before eleven."

"A misty morning may have a fine day." "Cloudy mornings turn to

clear evenings." "Rain before seven, clear before eleven." "If rain begins at early morning light, 'twill end ere day, at noon is bright." (English).

"Morning rains are soon past." (French).

"When it rains in the morning, it will be fine at night." "When it rains about the break of day, the traveller's sorrows pass away." (Chinese).

"Three foggy or misty mornings indicate rain." (American: Western U. S.)

A flood in the river means fine weather. (Welsh).
"A river flood, fishes good." (Spanish).

After a rainy winter follows a fruitful spring. (English).

"If there is much rain in winter, the spring is generally dry." (Greek).

*"Rain in September is good for the farmer, but poison to the vin**e** growers."* (German)

After clouds a clear sun.
(Latin).

"After clouds clear weather."
"A southerly wind and a cloudy
sky proclaim it a hunting
morning."

"When clouds after rain
disperse during the night, the
weather will not remain clear."

"Cloudy mornings turn to clear
evenings."

"When the clouds of the morn to
the west fly away, you may
conclude on a settled fair day."
"If clouds be bright, 'twill clear
tonight; if clouds be dark, 'twill
rain, do you hark?"

"If the sky beyond the clouds is
blue, be glad, there's a picnic
for you." (English).

After rain comes heat. (Welsh).

A green Christmas makes a fat
churchyard. (English, Scotch,
Danish).

"A black Christmas makes a fat churchyard."

"Many slones, many groans." When there is abundant fruit on the black thorn, there will follow a hard winter with much poverty and suffering.

"Many nits, many pits." When the nut trees are full of nuts, one may expect a large number of deaths and burials.

"When roses and violets flourish in autumn, it is an evil sign of plague and pestilence during the following year." (English).

John Ray, commenting on this proverb, declared that there was no great mortality nor epidemic in England during the summer and autumn of 1667, yet the preceding winter was unusually mild and that the last great plague that visited the country followed a very severe and frosty winter.

A mackerel sky never holds three

days dry. (English).

 "Mackerel sky, mackerel sky, never long wet and never long dry."

"Mackerel clouds in sky, expect more wet than dry."

"A mackerel sky is as much for wet as 'tis for dry."

"Mackerel scales, furl your sails."

"A mackerel sky, not twenty-four hours dry."

"A mackerel sky denotes fair weather for that day, but rain a day or two after."

"Mackerel sky and mares' tails make lofty ships carry low sails." (English).

 "It is belief even among educated people that what is called a mackerel sky prognosticates wet. In Scotland they hold the same thing of the clouds when they present three distinct shades.

In Carr's Dialect of Craven, 1828, i., 221, it is said that Hen Scrattins are 'small and circular white clouds denoting rain or wind. A friend informs me,' says the writer, 'that it is usual in Devonshire for the people to say, "See mackerel backs and horse-tails," as indicative of rain or wind.'"—C. Carew Hazlitt.

A March wisher is never a good fisher. (English, Scotch).

March, when blustering and stormy, is not a good month for fishing.

An evening red and a morning grey, two sure signs of one fair day. (English).
  See Matt. xvi:2, 3.

 "An evening grey and a morning red will send a shepherd wet to bed."

"Evening grey and morning red make the shepherd hang his head."

"Evening grey and morning red,

put on your hat or you'll wet your head."

"A red evening and a white morning, rejoice the pilgrim." (English).

"A red evening and a grey morning set the pilgrim a-walking." (Italian).

"An evening red and morning grey make the pilgrim sing." (French).

"Evening red and weather fine; morning red, of rains a sign." (German).

"The evening red and morning grey are the tokens of a bonnie day." (Scotch).

"A red sky in the morning, occasional showers; a red sky in the evening, fine weather is ours." (Welsh).

A rainbow in the morn, put your hook in the corn; a rainbow at eve, put your hook in the sheave. (English).

"If the rainbow comes at night, the rain has gone quite."

"A rainbow in the morning is the shepherd's warning; a rainbow at night is the shepherd's delight." (English).

This last proverb is sometimes given in the following rhyme:

"The rainbow in the morning
Is the shepherd's warning
To carry his coat on his back;
The rainbow at night
Is the shepherd's delight
For then no coat will he lack."

"Rainbow to windward, foul fall the day; rainbow to leeward, damp runs away." (English).
"Rainbows with the new moon, rain until the end." (Welsh)

"The rainbow has but a bad character, she ever commands the rain to cease."

"If there's a rainbow at eve, it will rain and leave."

"The boding shepherd heaves a sigh, for see, a rainbow spans the sky."

"When rainbow does not touch water, clear weather will follow." (American).

"If the rainbow appears when the rain has just begun, the earth will be filled; if at the end, it is a sign that the rain will stop." (Behar).

"The weather's taking up now
 For yonder's the weather gaw;
How bonny is the east now!
Now the colors fade awa'."
                    —Scotch Rhyme.

The weather gaw—i.e., a fragment of a rainbow.

"A weather-gall at morn, fine weather all gone;
A rainbow towards night, fair weather in sight.
Rainbow at night, sailor's delight;

Rainbow in morning, sailors
take warning." (English).

If the partridge sings when the
Rainbow Spans the sky,
There is no better sign of wet.

At twelfth day, the days are
lengthened a cock's stride.
(Italian).

"Some say that, if on the twelfth
of January the sun shine, it
foreshows much wind. Others
predict by St. Paul's Day
(January 25th), saying if the
sun shine it betokens a good
year; if it rain or snow,
indifferent; if misty, it predicts
great dearth; if it thunder, great
winds and death of people that
year."
—*Shepherd's Almanac* (1676).

At twelfth day, the days are
lengthened a cock's stride.
(Italian).

Some say that, If on 12 January the
the sunshine it for Seattle's much

wind. Others predict that St. Paul's
day (January 25), saying if the
sunshine it be tokens a good
year. If it rain or snow, in different;
it misty, it predicts great
dearth: if it's thunder, great winds
and death of people that year.
(Spanish almanac. 1676.)

Better be bitten by a snake, then to
feel the son in March.(English).

March flowers make no summer
Bowers. March damp and warm
will do no farmer much higher
(English).

A dry march never brings it bread.
March grass never did good
(American).

When flies swarmed in March,
sheep come to their death. When
Nance dance in March should
brings death to sheep (Dutch).

The March sun wounds.(Spanish).

Better slaughter in the country in
March should come in mild.

(Manx).

Bullions day, gif ye be fair, for 40 days 'twill rain nae mair. (Scotch).

St. Martin Bullion's, July 4. It Bullion day be dry, there will be a good harvest. If the deer rise dry and lie down dry on Bullion day, there'll be a good gose harvest. "Gose" refers to the latter part of summer (Scotch).

Comets bring cold weather. (English).

In France, comets are thought to improve the grape crop, and wine that is made during the year of their appearance is called "comet wine").

Expect not fair weather in winter From one night's ice. (English).

Good signs of rain't always he'p de young crops (America).

Hail brings frost from with its tail. (English).

Hark! I hear the asses bray, we
 shall have some rain today.
(English).

Hen scarts and filly tails make
Lofty ships wear low sails.
(English, Scotch).

If clouds look as if scratched by
A hen, get ready to reef your
 Topsails then. (English sailors'
proverb).

If cold at St. Peter's Day, it will
last longer (English).

It is also said that, "The night of
February St. Perer's (February 22nd)
shows what the weather will be for
the next forty days" (English).

If it rains before seven, 'twill cease
Before eleven (English).

The following weather signs are
held by some to be trustworthy:
If it rains before daybreak it will
cease before eight o'clock in the
morning.

If it rains before the sun shines, it
will rain the next day.

If it rains between eight and nine
o'clock in the morning, it will rain
till noon.

If rains begin about noon, it will
Continue through the afternoon.

If rain begins after nine o'clock
in the evening, it will rain the
next day.

If rain begins an hour before
daybreak, it will probably rain
all day.

If rain begins about five o'clock
in the evening, it will rain all night.

If rain comes after midnight, it will
rain the next day.

If rain ceases before midnight, it
will be clear the next day.

If rain does not cease before noon,
It will continue till evening.

If red sun begins his race, expect

that rain will flow apace (English).

A red sun has water in his eye. (English).

The side being red at evening
forshewes a faire and cleare
morning:
but if the morning riseth red
of wind and raine we shall be
sped. (A. Fleming).

If robins are near houses, it is a
sign of rain. (English).

If the robin sings in the bush,
then the weather will be course.
If the robin sings on the barn,
Then the weather will be warm
(Old English Rhyme).

If the cock drink in summer it will
rain a little after (Italian).

Cocks are said to clap their wings
in an unusual way, and to crow
more than usual and at an earlier
hour, just before rain (English).

If the cock goes crowing to bed,
he'll certainly rise with a water

head (English).

If the cock moult before the hen,
We shall have weather thick and
 thin; but if the hen moult before
the cock, we shall have weather as
hard as a block.
(Old English Rhyme).

If the crow speak by night and the
jackal by day, there will be either
a rain storm or an inundation
(Behar).

If the first three days of April be
foggy, rain in June will make the
lanes bogy. (English).

If the first thunder is from the east
 the winter is over. (Zuni Indians).

"After the first thunder comes the
rain." "If the first thunder is in the
east, aha! the bear has stretched
his right arm forth, and the winter
is over." "With the first thunder
the gods rain upon the petals." "If
the first thunder is in the south,
 aha! the bear has stretched his
right leg in his winter bed." "If
the first thunder is in the west,

ha! the bear has stretched his left
arm in his winter bed."

"When the clouds rise in terraces
of white, soon will the country of
the corn priests be pierced with
 the arrows of rain."

 "With the rain of the
north-east comes the ice fruit"—
hail.

"When frogs warble, they
herald rain." "The west rain comes
 from the world of waters to
moisten the home of the She Wi."
"The moon, her face if red be, of
water speaks she."

"When the butterfly comes, comes
 also the summer."

"When the dew is seen shining on
the leaves, the mistrolled down from
 the mountains last night."

"When the sun sets sadly, the morning
 will be angry."

"When the sun is in his house
(surrounded by a halo), it will rain

soon."

"The moon if in house be,
cloud it will, rain soon will come."
—*Zuni Indian Weather Sayings*
(U.S. Signal Service Notes IX.
Weather Proverbs).

If the halo is seen round the moon
on Sunday (night), it will rain the
day following; if on Thursday,
(it will rain) the day following;
and if on Tuesday, (it will rain) on
the eighth day. (Behar).

"Far burr (halo), near rain; near
burr, far rain." "Bigger the ring,
nearer the wet."

"The moon with a circle brings
water in her beak."

"A lunar halo indicates rain, and
the number of stars enclosed,
the number of days of rain."

"When the wheel is far, the storm
is n'ar; when the wheel is near, the
storm is far." (English).

"When round the moon there is a

brugh (halo), the weather will be cold and rough." "A far brugh, a near storm." (Scotch).

"Circle near, water far; circle far, water near." (Italian).

"A halo round the moon is a sign of wind." (Chinese).

If the oak's before the ash, then you'll only get a splash; if the ash precedes the oak, then you may expect a soak. (English and Scotch).

It is a common belief that one can tell whether the summer will be dry or wet by the leafing of the trees.

Another English saying asserts that "If the oak is out before the ash, 'twill be a summer of wet and splash; but if the ash is before the oak, 'twill be a summer of fire and smoke"—which has been abbreviated by the Kentish folk to "Oak smoke, ash squash."

Other forms of the saying are found in different parts of

England and Scotland. The only proverb related to the above that can be relied upon is used in Surrey where the people say, "If the oak before the ash come out, there has been or there will be drought."

If the Pleiades rise fine, they set rainy. And if they rise wet they set fine (Swahilian).

If there be neither snow nor Rain, then will be dear all sorts of grain (English).

If there is ice in November That will bear a duck, there will be nothing after but sludge and muck (English).

Ice in November brings mud in December. If the ice will bear a goose before Christmas, it will not bear a duck after.

If the geese at St. Martin's Day (November 11) stand on ice, they will walk in mud on Christmas (English).

If ducks do slide at Hollantide,
at Christmas they will swim.
If ducks do swim at Hollantide,
at Christmas they will slide.

If you see cloudless nights
and cloudy day, be sure, says
Ghagh, that the rains are at
an end (Behar).

In the wane of the moon, a
Cloudy morning bode a fair
afternoon (English).

It is better to see a troop of
Wolves than a fine February.
(French).

Warm February, bad hay crop.
Cold February, good hay crop.
All the months in the year curse
a fair February.

The Welshman had rather see his
dam on the bier, than to see a fair
February.

A February singing never stints
stinging.

A February spring is not worth a
pin.

February fill the dyke, weather
either black or white. but if it be
white, it's better to like.

In February if thou hearest
thunder, thou wilt see a summer'
Wonder (English).

One would rather see a wolf in
February than a peasant in his
shirt sleeves (German).

If in February there be no rain,
'tis neither good for hay nor

Grain (Spain, Portugal).

February rain is only good to fill ditches.

February rain is as good as manure.
Snow in February puts little wheat in the granary (French).

Snow which falls in the month of February puts the usurer in a good humour (Italian).

When it rains in February it will be temperate all year (Spanish).

When February gives snow, it fine weather foreshows (Norman French).

It will be the same weather for nine weeks as it is on the ninth day after Christmas. (Swedish).

March dry, good rye; March wet, good wheat. (English).

"March rainy, April windy, and then June will come

beautiful with flowers."
(Spanish).

"A dry March, wet April, and
cool May, fill barn, cellar,
and bring much hay."
(English).

Mist in spring is worse than
 poison. (Welsh).
 "Mist in spring is a sign of snow." "Mist
in summer is a sign of heat." "Mist in
autumn is a sign of rain." "Mist in winter
is a sign of snow." (Welsh).

North-west is far the best,
north-east is bad for man and
beast. (English).

There are a vast number of
proverbial sayings about wind
and weather; a few only are here given:
"Look not, like the
Dutchman, to leeward for fine weather."

"Wind roaring in chimney, rain to
come."

 "A veering wind, fair weather; a backing

wind, foul weather."

"If the wind be hushed with
sudden heat, expect heavy
rain."

"A high wind prevents frost."

"A northern air brings weather
fair."

"Do business with men when the wind is
in the north-west."

"When the wind is from the east,
it is four and twenty hours at
least."

"An easterly wind's rain makes
fools fain."
"The wind in the West suits
everyone best."
"Wind west, rain's nest."
"When wind is west, health is
best."

"A western wind carrieth water
in his hand." (English).

"No weather ill, if the wind be still."

(English and Scotch).

"A west wind, north about, never hangs lang out." (Scotch).

"A north wind has no corn." (Spanish).

"Great heat brings wind." "The east wind breaks up the frost." (Chinese).

"A north wind with new moon will hold until the full." (American).

"North wind show de cracks in de house." (American).

"If the east wind blows in Sāwan (July and August), sell your bullocks and buy cows."

There will be no ploughing. "If the west wind blow in Sāwan for only two or three days, rice will grow even behind your hearth."

"When the wind blows from all quarters, there is hope of rain." (Behar).

The following Zuni Indian
sayings, as given in the *Notes of
the United States Signal Service*, Note
IX., will be of interest:

"Wind from the North, cold and snow.
Wind from the Western river of
the Northland (Northwest wind), snow.
Wind from the world of waters
(West wind), clouds.

Wind from the Southern river
of the world of waters
(South-west wind), rain.

Wind from the land of the
beautiful red (South wind),
lovely odours and rain.

Wind from the wooden cañons (South-
east wind), rain and
moist clouds.

Wind from the land of day, it is
the breath of health and brings
the days of long life.

Winds from the lands of cold
(North-east wind), the rain
before which flees the harvest.

Winds from the lands of cold
(North-east wind), the fruit
 of ice.

Wind from the right hand of
the West is the breath of the
 God of Sand Clouds."
        "The west wind always brings wet
        weather,
        The east wind cold and wet
        together,
        The south wind surely brings
        us rain,
        The north wind blows it back
        again."
                        —*Old English Rhyme*

"When the wind is in the
East, then the fishes bite the least;
When the wind is in the
West, then the fishes bite the best;
When the wind is in the
North, then the fishes do come
forth;
When the wind is in the
South, it blows the bait in
 the fish's mouth."
                        —*Old English Rhyme.*

"When the wind is in the
North, hail comes forth;
When the wind is in the
West, look for wat blast;
When the wind's in the
South, the weather will be
fresh and gude;
When the wind is in the
East, cauld and snow comes meist."
Old Scotch Rhyme.

"Winter
"North winds send hail,
South winds bring rain,

East winds we bewail,
West winds blow amain;
North-east is too cold,
South-east not too warm,
North-west is too bold,
South-west does no harm.

Spring
The North is a noyer to grass
of all suits;
The East a destroyer to herb
and all fruits.

Summer
The South, with his

showers, refresheth the corn;
The West to all flowers may
 not be forlorne.

Autumn
The West, as a father, all goodness
doth bring;
The East, a forbearer, no
manner of things;
The South, as unkind,
draweth sickness too near;
The North, as a friend,
maketh all again clear.

With temperate wind, we
blessed be of God,
With tempest we find we are beat
with His rod;
All power, we know, to
remain in His hand,
However wind blow, by sea
or by land."
              —Thomas Tusser.

On St. Michaelmas Day the devil puts his
foot on the blackberries. (Irish).

St. Michaelmas Day, September 29th.

On St. Barnabas's Day the sun

comes to stay. (Spanish).

St. Barnabas's Day, June 11th.

Rain before church, rain all the week little or much. (English).

"If there is rain in the Mass, 'twill rain through the week either mair or less." (Scotch).

Rain in Chitra (October) destroys the fertility of the soil and is likely to produce blight. (Behar).

Saturday's new, and Sunday's full was never fine, and never wool. (English).

"If the moon change on a Sunday there will be a flood before the month is out."

"A Saturday moon if it comes once in seven years, comes once too soon." (English).

"A Wednesday's change is bad." (Italian).

"Saturday's moon and Sunday's prime, once is enough in seven years' time." (Scotch).

"If the weather on the sixth day is the same as that on the fourth day of the moon, the same weather will continue during the whole moon." (Spanish).

So far as the sun shines on Christmas Day, so far will the snow blow in May. (German).

"If the sun shine through the apple tree on Christmas Day, there will be an abundant crop in the following year." (English).

St. Mamertius, St. Pancras, and St. Gervais do not pass without frost. (French).

That is, frost is sure to come on May the eleventh, twelfth, or thirteenth.

The barking of the fox and the flowering of the kās grass are signs of the end of the rains. (Behar).

"The appearance of the star Canopus and the flowering of

the kās grass in the forests are
signs of the end of the rains."

"The kās grass and the kus grass
flower on the fourth of the light
half of Bhad (August and
September), why do you plant
out, O cultivator!" for there will
be no more rain. (Behar).

The dirt bird sings, we shall have
rain. (English).

The dirt bird—*i.e.*, the dirt owl.

The screeching of an owl
indicates cold or storm.
The hooting of an owl at night
indicates fair weather.
The crying of an owl in storm
indicates fair weather.
The crying of an owl in fair
weather indicates storm.
The screaming of an owl in bad
weather indicates change of
weather.
> —*Old Weather Signs.*

In Syria the owl is called the
"Mother of Ruins"; in China,

the "Bird which Calls for the Soul"; in Ireland, the "Old Woman of the Night."

The first three days in January rule the coming three months. (English).

"The month of January is like a gentleman": As he begins so he goes on. (Spanish).

"A favourable January brings us a good year." (English).

The full moon brings fair weather. (English).

"The full moon eats clouds."
"The moon grows fat on clouds."
"Near full moon a misty sunrise bodes fair weather and cloudless skies." "If the full moon rise red expect wind."

Thunder in spring, cold will bring. (English).

"Early thunder, early spring."
"Lightning in summer indicates good healthy weather."

"Thunder in the fall indicates a mild open winter." "Winter thunder bodes summer's hunger." (English).

January thunder indicates wind, corn, and cattle.
February thunder indicates poor maple-sugar year.
March thunder indicates coming sorrow.
In Germany thunder in March is thought to indicate a fruitful year.
April thunder indicates a good hay and corn crop.
May thunder indicates that there will be no thunder during August and September.
July thunder indicates that the wheat and barley will suffer harm.
August thunder indications do not come alone: one thunder storm will follow another.
September thunder indicates a good crop of grain and fruit.
In Germany, thunder in September is thought to indicate snow in February and March and a large crop of grapes.
November thunder indicates the

coming year will be fertile.
December thunder indicates good
weather.

Old English Weather Signs

Ughun is water on the fire
(Hindustani).
September and October (Coar)
is but the gate of gold.
October and November (Cartic)
ends, yet scarcely told.
November and December (Ughun)
just lets water seethe.
December and January (Poos)
makes us but in corners breathe.
January and February (Magh)
lengthens by minute degrees.
But February and March (P'hagun)
Straightens our knees:
Then March and April (Cheyt) the
pleasant year replaces and dirty
fellows wash their faces.

By the time it takes to boil water
does the day lengthen.

When February gives snow, it fine
Weather foreshows (Norman)

When small water snakes leave the

sand in low damp lands, frost may be expected on three days (Apache Indians).

When the cat lies on its brain, it is Going to rain (English). Lies on its brain means lies on his back.

When a cat sneezes, it is a sign of rain.

When a cat scratches the table legs, A change in weather is coming.

If the cat washes her face o'er the ear, 'tis a sign the weather 'ill be fine and clear.

When cats wipe their jaws with their feet, it is a sign of rain.

The cardinal point to which a cat turns and waxes her face after a rain shows the direction from which the wind will blow.

The old woman promised a fine day on the morrow, because the cat's skin looked bright.

When a cat scratches itself, or scratches on a log or tree, it Indicates rain.

When sparks are seen on stroking a cat's back, expect a change in weather.

When a cat washes its face with its back to the fire, expect a thaw in winter (English).

When the cat lies in the sun in February, she will creep behind the stove in March (English, German).

Cats wash their faces before a thaw.

Cats sit with their backs to the fire before snow.

Cats scratch a wall or post before Wind (Scotch).

Putting a cat under a pot brings bad weather (Irish).

When the cat turns towards the North and licks its face, the wind

Will soon blow from that direction (Greek).

When the clouds fly like the wings of the partridge and when a widow smiles, one is going to rain and the other to marry. (Behar).

When the days begin to lengthen, the cold begins to strengthen. (English).

"As the days begin to shorten, the heat begins to scorch them." (English).

When there is thunder rain falls. (Marathi).

DIY

DO-IT-YOURSELF

Here are some interesting projects one may like to do. Some of these projects, one could just go by the instrument but it is more fun to make the instrument and get a better understanding as to how the weather is affecting that instrument.

These projects are especially fun with kids and even done as science projects for school.

MAKE A WEATHER BAROMETER

Cut the top off a two-liter plastic bottle so that the curve of the bottle is cut away leaving straight sides.

Set a ruler inside the bottle so it is standing up flat against the side of the bottle. Keep the ruler to the edge of the bottle so that the numbers are visible.

Tape about 16 inches of clear tubing alongside the ruler so that the bottom of the tubing is slightly above the bottom of the bottle.

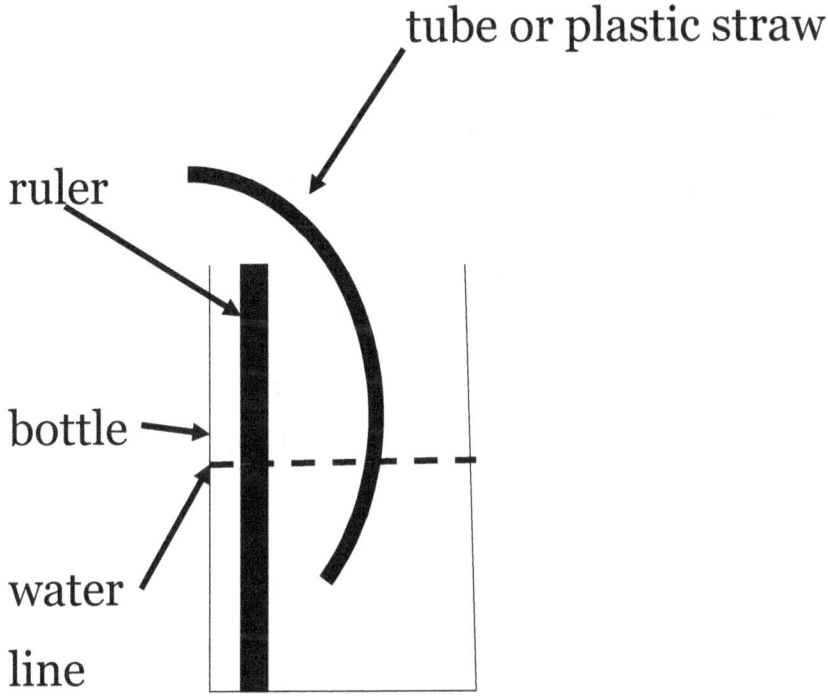

tube or plastic straw

ruler

bottle

water

line

Fill the bottle halfway with water. If you like you can add a little coloring or food dye.

Like using a straw suck a little water up the tube until it is well above the water level in the bottle. Now the trick is to seal the top of the tube so that the water does not recede down the tube. Some have used a piece of chewing gum to seal the tube.

The water level in the tube must be within range of the ruler for measurement. Mark this line on the tube in the water line inside the bottle.

As a barometer the water in the tubing should rise when the weather is clear and to fall when it is cloudy or rainy.

Remember a barometer, just like this one, does not predict weather it only predicts air pressure which gives an indication as to possible weather. High-pressure good weather: low-pressure possibly bad weather.

Best results are obtained if the barometer is outside. A barometer indoors shows the air pressure indoors.

FINDING TRUE NORTH

Northern Hemisphere with watch compass)

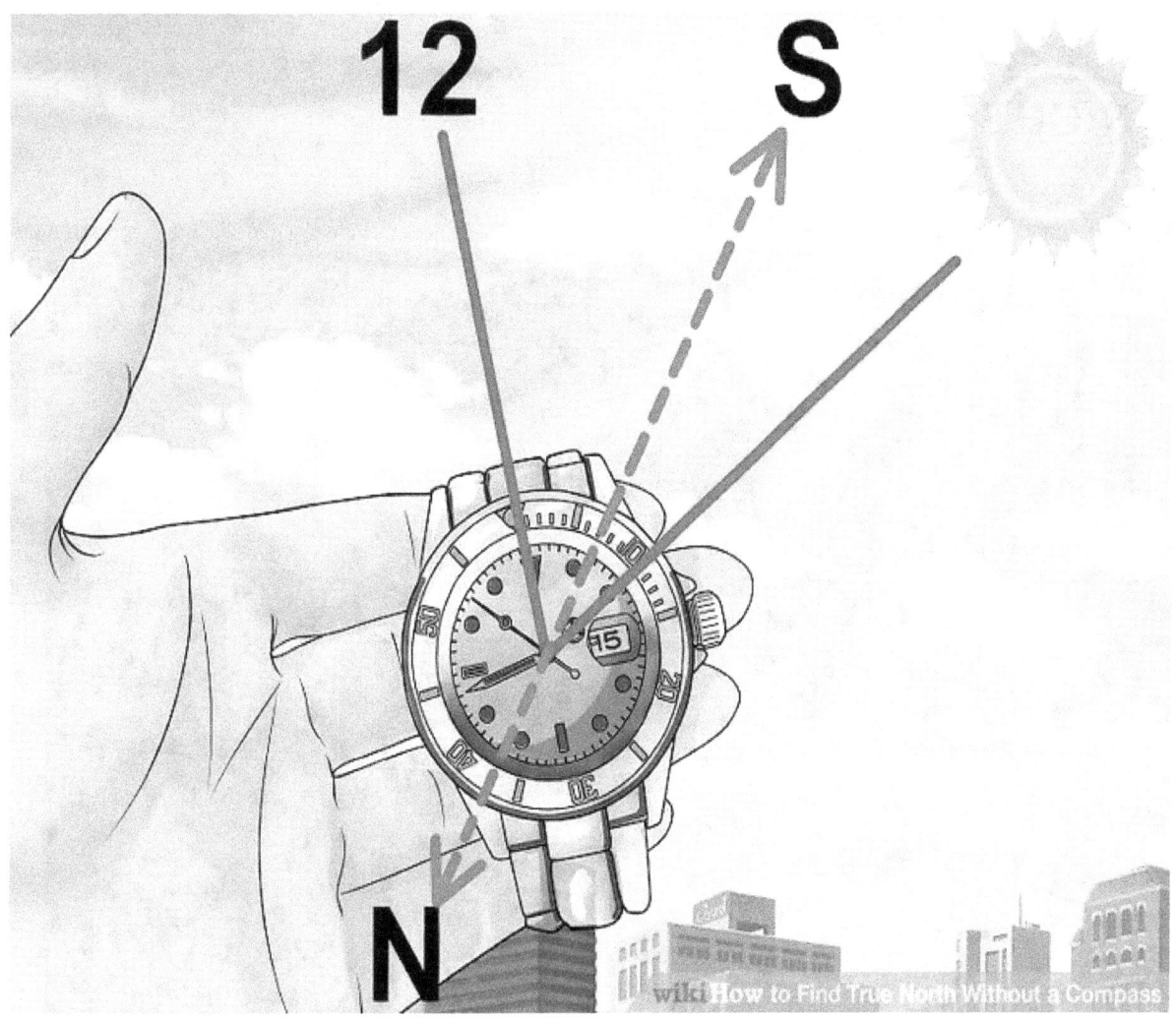

With an analog watch (it has hands or arms for minutes and hours). Hold the watch horizontal to the earth. Point the hour hand at the sun. Form an angle from the 12 o'clock mark to the hour hand. Bisect this angle for the north-south line. Since the sun rises in the east and sets in the west you should barely determine which is south.

AT NIGHT

If it is night in the northern hemisphere locate the Northstar (Polaris) which is the last star in the handle of the Little dipper. Draw a line straight down from the Northstar to the ground and that is North.

If you have trouble finding the Little Dipper, locate the Big Dipper and the two stars of the Big Dipper will point to the Northstar.

FINDING NORTH WITH OUT A COMPASS MAGNET OR SUN

If possible, look for a spider web. Spiders spin their webs to face true south.

Moss generally grows on the southward side of a tree.

TIME LEFT BEFORE SUNSET

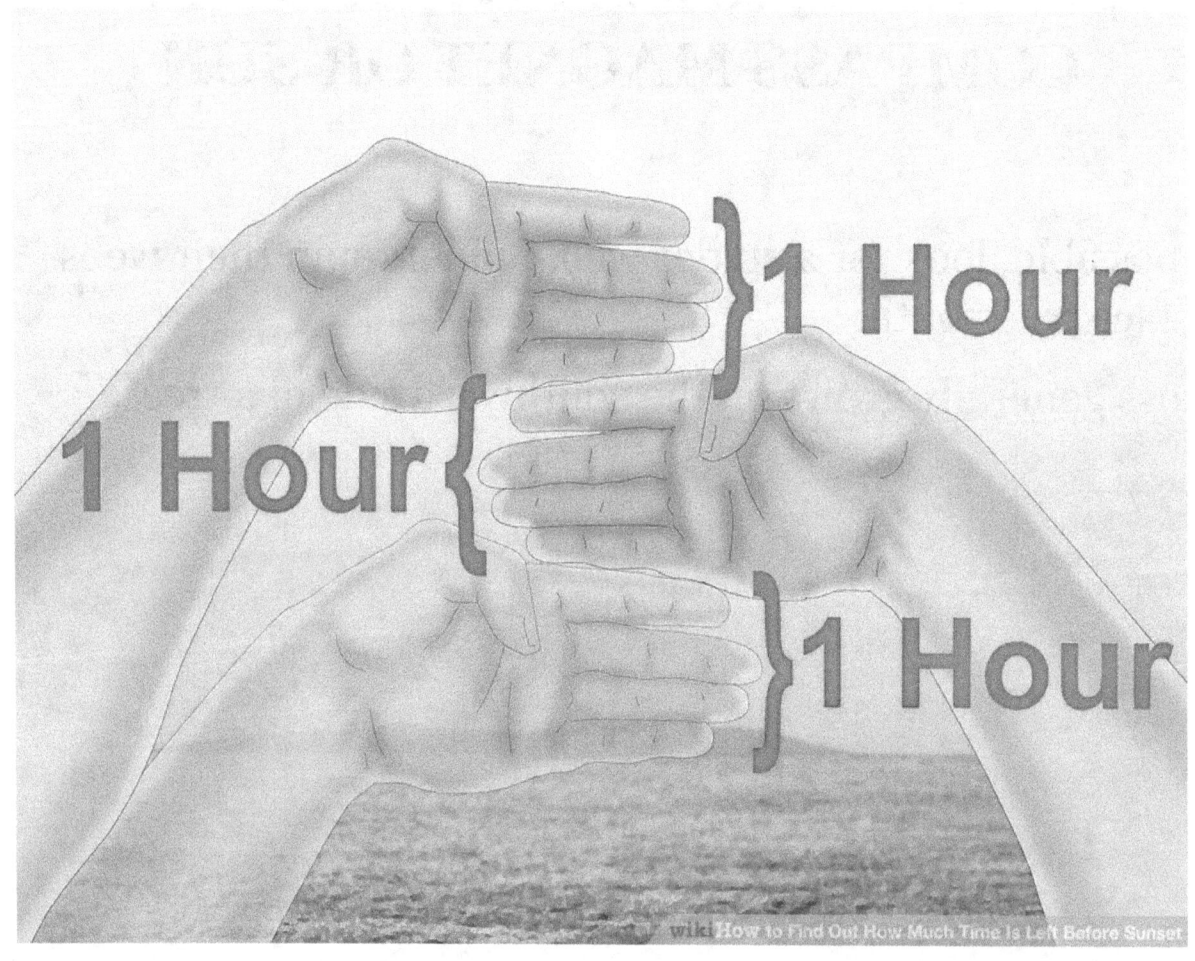

The diagram pretty well explains it.

HIGH-PRESSURE LOW-PRESSURE

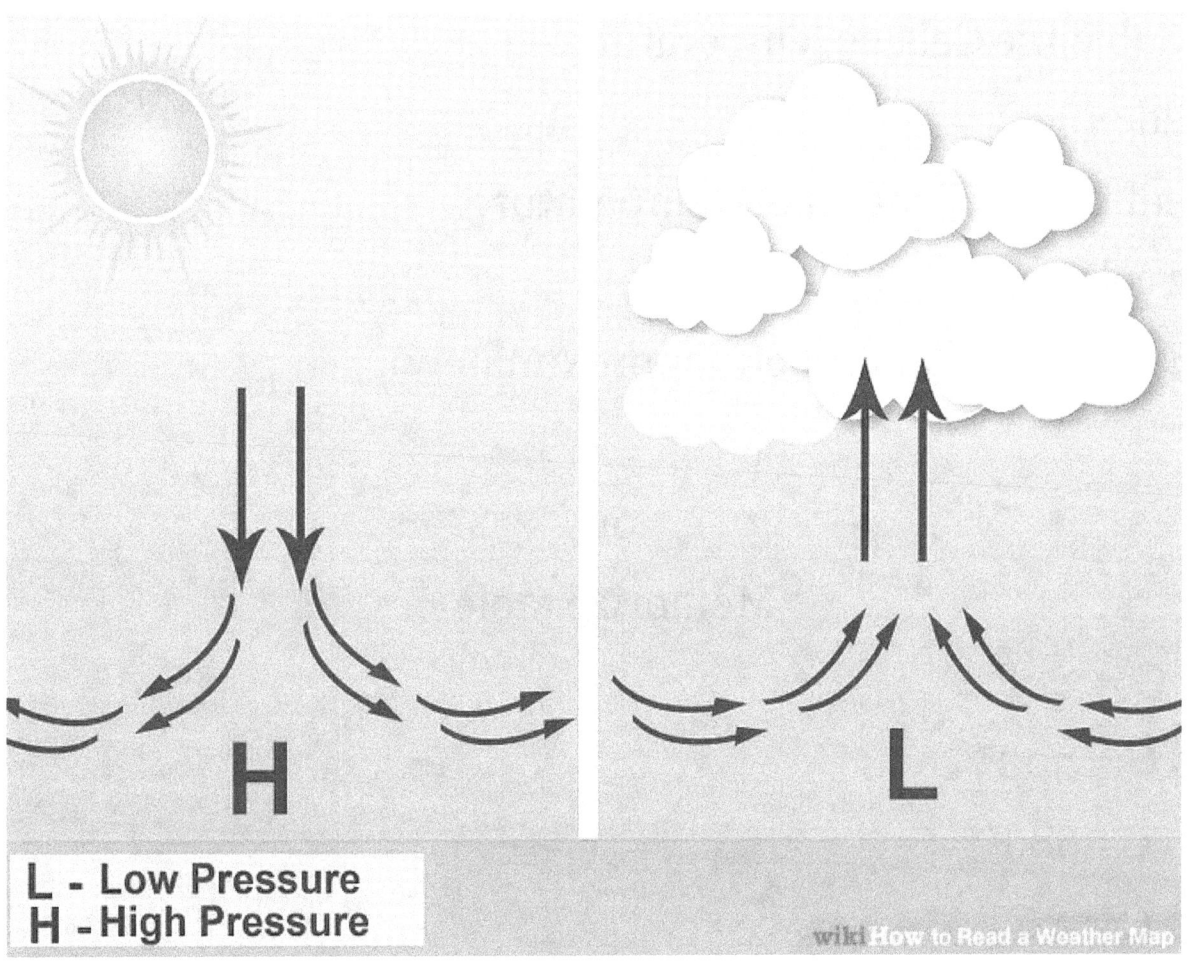

L - Low Pressure
H - High Pressure

As the diagram shows:

High-pressure is dense air falling to the ground to create clear whether. Low-pressure air is less dense or lighter as a result it rises and cools to form clouds and possible rain.

ESTIMATE CELSIUS TEMPERATURE TO FAHRENHEIT

Double the Celsius temperature.

Add 30

That is your Fahrenheit temperature.

22 celcius X 2 =44.

44 + 30 = 74 fahrenheit (approximately).

or

Memorize table

0 C = 32 F

10 C = 50 F

20 C =68 F

30 C = 86 F

40 C = 104 F

SUMMARY

These weather predictions and forecasts are interesting, but the truth and the accuracy may lie in your interpretation and your experimentation and observations as a test of their validity.

Many of these sayings do have a scientific basis but even with the high tech scientific equipment, predictions are sometimes wrong because weather changes quickly and surprisingly.

Whatever you decide, it is fun to make a prediction and see if is accurate.